W9-AOR-464

# EYES

## WINDOWS OF THE BODY AND THE SOUL

LA DEAN GRIFFIN

# EYES

## WINDOWS OF THE BODY AND THE SOUL

BiWorld

First Printing
December, 1976

Second Printing
September, 1977

ISBN 0-89557-007-6



Printed in the United States of America
by Microlith Printing

# DEDICATION

To my father James Melvin Gibby whose philosophy I have added to my own in this book, with love and gratitude for all the truth he taught me and for the example he so lovingly and faithfully set for me.

## PREFACE

To write a book on iridology and try to surpass the grand work of Dr. Bernard Jensen would indeed be presumptuous on my part. Dr. Jensen has preserved and popularized iridology for many years with a dedication and love for people, despite heavy pressures, that has been an inspiration to all of us. My purpose in writing this book is merely to add another witness to his work.

In addition, I have learned some things not only about iridology but how to relate iridology with the use of herbs and a mild food diet. A person cannot only recognize improvement beause of the way he feels but he can also watch the changes as they take place in the eyes. This is a marvel and a wonder to behold, for as the body heals, the eyes change like a kaleidoscope. If such results can be achieved, the logic of a mild food diet and use of herbs must be sound, and iridology must then be a Truth.

The use of drugs for healing by established medical methods has too long proven ineffective. It has been previously assumed that drugs were all we needed, and in the beginning the whole idea of finding a wonder drug to solve the world's physical problems met with little resistance. Everyone liked the idea, for it took no effort merely to take a pill and settle all problems. However, thinking people everywhere in the world today have become disenchanted with this idea and the drugs themselves. Mankind is coming to the conclusion that drugs are alien to the natural physiology of man. Medical science's quest has resulted in failure, for in that quest all the rules and physical laws of the body have been broken. Those who are thinking of looking for alternative solutions are coining a new word — biological medicine — and returning to the herb kingdom for the answers. However, those who are thinking even deeper are recognizing that the real answer does not lie in medicine of any kind but rather in obedience to the laws of nature. Only then will disease flee from us.

# IRIDOLOGY

Iridology or iro-diagnosis had its roots centuries ago in ancient middle-east medicine. It has never reached, as far as I can determine, its full potential as a valuable diagnostic tool in the hands of skilled iridologists. Modern medicine with its marvelous optical systems developed the ability to peer through the pupil, revealing the intricate mysteries of the retina, while at the same time by-passing the external, more readily accessible iris. Clinical practitioners today are generally quite ignorant of iridology except for one sign: the arcus senilis. Yet if the doctor is questioned (How is it that systemis hardening of the arteries shows up in the iris of the eye? How is the iris hooked up, neurophysiologically speaking, to the heart and blood vessels in the limbs?) A blank stare is the usual response. Iridology is not taught in schools of medicine in the U. S. but is studied, however, in some foreign schools, particularly in Germany.

The science of iridology is based upon very precise anatomical and physiological mechanisms which our creator has designed for our use and benefit. Not all are yet ready to accept the concept. But for those who have, iridology has become a wonderful blessing in their lives since these eye-signs are often the first objective changes indicative of a disease process. (You should be aware that pigeon breeders have been depending on eye-signs in their prize racers for many years.)

I am so pleased to write this introduction to this important, new book by my dear friend LaDean Griffin, a most sincere and enlightened lady. LaDean's purpose in her writing and lecturing is to share her knowledge and understanding with her brothers and sisters everywhere, so they might help themselves and their loved ones in maintaining rich, abundant health and promoting healing through natural methods. LaDean has soul and she teaches from the heart, advocating all safe therapies which help our natural abilities in health considerations. Thank you LaDean for a most provocative and challenging volume. Many will call you blessed because of it.

Robert Bliss Vance D. O.

Salt Lake City
June 2, 1976

# CONTENTS

# CHAPTER ONE

# A LOOK INSIDE

*Pitifully Poor Politics*

*It's pitifully poor politics when pretentious politicians and pilfering private promoters purposely prevent or postpone public protection from possible pillage and potential plunder while they painstakingly perpetrate plans for personal plutocracy . . . What a painful preposterous price the people pay for such pinheaded pigishness. - - - and, as long as we tolerate such black and blighting blunders we'll continue to be in the* red. *So, for* Freedom's Sake *and our own, let's be* Real, *United Americans in Our United States.*

J. Melvin Gibby
1964

**Think**
**America First!**

Aside from the fact that people get better on my programs, of all the subjects I teach, iridology is probably the study which proves my theories more than any other single thing. As people watch the change for the better in their general health, they are amazed to watch the changes taking place in the iris of their eyes. It is both a thrilling and almost magical transformation that occurs: an exciting new demonstration of the art of healing — or at least it is new to modern man. Iridology: the practice of analyzing the inside of the body by the iris of the eyes. It dates back to the 12th Century, and has probably been known by some great physicians as far back as the beginning of time.

My purpose is to teach some of the main things to look for and how to use herbs, mild foods and vitamins in conjunction with what is discovered in the eyes. The study of iridology could be an important facet of your life. We have been presented with a grave spectacle of physical degeneration increasing daily around us as a result of fear tactics through the wide use of drugs. It has not been without reason that these dire results have caused such general dependence upon the Doctor.

As we have watched the jubilation over the wonder drugs sink into the decay of degenerative disease they cause, we can see that to try to find the answers ourselves, when it is almost too late, is a lonely ordeal.

Reading through an old book on iridology dated 1924, I suddenly thought that the medical doctor who wrote it had never in his wide experience ever seen the kind of eyes we see today. The eyes we see today since 1924 and the 50-year use of drug therapy, look shattered, filled with all manner of degeneration not known in 1924. We even see in the photos taken in my classes the most seriously degenerative diseases of old age showing upon the eyes of babies and young children. Another fifty years of drugs and we will not have to worry about the population explosion. On the contrary, we will have to worry about taking care of all the blithering idiots and sick old babies we will create in future generations. We now have more retarded or maimed children or children with birth defects than ever before in history. The proportion of Americans wearing glasses has also increased. Then we have the audacity to proclaim the great advances of medical science. Thinking people everywhere are searching for alternative solutions to their health problems and finding them among the natural, simple and inexpensive medicines God has provided in the herb kingdom. Medicine rip-offs by doctors and pharmacists are costing the U. S. taxpayers three billion dollars a year, says a recent newspaper. But what about the rip-off that inevitably occurs through our being a drug-oriented society? What about the drug-caused diseases we contend with today? An article came out in a Los Angeles paper by a large medical hospital which is trying to overcome the problems of errors made in giving medication. It said that the average doctor is only familiar with a tiny percentage of the 7000 principal drugs in the medicinal arsenal. As a result wrong dosages and wrong drugs are used. It was stated that 100,000 serious errors are made daily in American hospitals in preparing, prescribing and dispensing medications. It was further stated that these mistakes are causing the deaths of thousands of hospital patients every year. Many patients are dying who don't need to die, and others are not recovering rapidly enough because of the errors in not receiving proper medication for their disease. Their reasoning was that the method to help correct the errors was close security programs to safeguard against nurses, doctors and pharmacists making these errors.

*Virchow said, "Some doctors make the same mistakes for 20 years and call that experience."*

Not only do we have the concern about misuse of drugs but we also have the concern about the drugs themselves. In addition, we have the dangers involved in the use of over 3,000 different chemicals being used in the distribution of commercial food products, many of which are suspected of being harmful. Some have been proven to be dangerous or at least suspect, says the Plain Truth magazine. Coal tar dyes are suspected of being harmful to man in nearly any amount if continually consumed. We see headlines warning us continually, such as, "Rat livers injured by color dye," "Cancer-inducing chemicals added to food for humans," Specialist says "Britain bans treated fruits from U. S."

Four to five hundred totally new chemicals are put in our food each year. What about the potential long-term hazards? Many articles are finding their way to the public about the poisons in our food and thinking people everywhere are becoming deeply concerned. Cyclamate alone has been contained in 172 brands of products with an annual sale of about a billion dollars. How many more poisons do we unwittingly consume?

Because of the controls of the FDA on drugs, natural herbal formulas are restricted by the medical monopoly on drugs, and yet herbs are not drugs per se. What about the FDA's control of all these new chemicals finding their way into our food supplies with no apparent restriction or control? Who is running this country: the scientists, the FDA or the people? A man who has been helping leper cases with herbs, with some startling results, was stopped by the FDA. He wrote to the President, dated November 17, 1973:

> Dear Mr. President:
>
> Today is the fourth anniversary of our being closed by the Food and Drug Administration. Although we

asked them to make an appointment for a demonstration, and showed them some two hundred or more letters from our clientel, my wife and I could see that they were not interested. They said our letters could be phoney. We asked them to select as many as they liked, and call them anywhere in the country on our telephone. He then stated, it is too late! I wanted to know why, as it is never too late to cure cancer. He said, now listen to me and to what I have to say, and I will only say it once, Mr. Crosby. From now on you are closed, and if you send or ship your product over state lines, you face IMPRISONMENT! At seventy years of age, it was an extremely difficult thing to accept, having your income suddenly stopped, without a chance to prove anything to anyone, (especially, after all the mistakes the FDA have recently made) is making a farce out of "FREEDOM OF FREE ENTERPRISE!"

Our formulae will cure cancer in any form, even tumor of the brain. It contains NO DRUGS, it is NON-TOXIC. Our formulae will cure MELANOMA completely in just fourteen days. When a patient has melanoma, the physicians tell the patient, there is no use for surgery, they have from 45 to 60 days left. This is not necessary, and we implore you sir, for an appointment, so that we may show you the efficacy of our formulae.

BREAST CANCER can be cured in sixty days without surgery, without drugs, chemicals or injections. Think sir, what this would do for WOMANHOOD, throughout the entire world!

Trusting we will hear from you in the near future, I remain sir,

Sincerely yours,
Kirley Crosby

This is the kind of answer Mr. Crosby receives for his inquiries:

May 4, 1971

Dear Mr. Crosby:

The White House has forwarded to us for reply your letter of March 11, 1971, concerning your formula for cancer.

If you propose to offer your formula for import into

this country, you should understand that it must comply with the Federal (U.S.A.) Food, Drug, and Cosmetic Act. Enclosed is a publication which described in non-technical terms the requirements of the Act.

We believe your formula may be a "new drug" as that term is defined by the Act. A "new drug" is a drug which is not generally recognized among qualified experts as safe and effective for its intended purposes. A "new drug" may not be distributed in interstate commerce in this country, or imported into this country, until its sponsor has submitted to us, and we have approved, a new-drug application containing substantial evidence (i.e., evidence based on adequate and well-controlled investigations, including clinical investigations) which establish that the drug is safe and effective for its intended uses.

The data to be submitted in support of a new-drug application must be based on the type of testing described in Regulations 130.3 and 130.4 of the enclosed copy of the New Drug Regulations.

Sincerely yours,
Donald N. Kilburn
Assistant to the Director
Division of Case Guidance
Office of Compliance
Bureau of Drugs

Pictured on page 63 is the leg before herbal treatment and after treatment.

There has to be a better way than such things being restricted by the powers that be. How soon will the people awaken and take back their rights from these powerful monopolies? The movement is on and we can see such a wave of health articles not known since the Vitamin craze of the thirties. People are taking a long hard look at the high cost of dying. Books are being written and movies pan the hospitals and doctors. People are studying and are not so willing to allow the doctor to rule them every waking moment. People are learning to take care of themselves and finding out that they can. The new thing is learning how to take care of yourself and keep yourself well. This is called preventive medicine, a popular new phrase that has been coined out of this struggle.

Many years ago Leah Widtsoe wrote:

*It is noble to help those who are ill and suffering, while trying to restore their health, but it is far nobler to teach one or a thousand how to build and maintain health so that suffering and disease may be prevented.*

In a letter to the Mormon Battalion dated August 19, 1846, Brigham Young said:

*If you are sick live by faith and let the surgeon's medicine alone if you want to live, using only such herbs and mild foods as are at your disposal.*[1]

The newspapers as far back as 1972 state that more than 3,000,000 persons needlessly undergo surgery each year, 36,000 resulting in death. On the other hand, a San Francisco paper says the organic food business has increased to such a phenomenal rate that there is now a far greater demand than supply of such fruits and vegetables. When world-wide health series are also beginning to emphasize disease prevention and articles such as those which came out in the "London Times" recently begin to agree with some new health ideas. But scholars and doctors will probably bull-doggedly hold onto tradition. As an example, a recent radio talk show interviewing such a person showed the sorry predicament of an information school-computed individual who had no ability to reason things out for himself. It seems that the young lady asking the questions was deliberately asking all the key questions against natural methods, using every few minutes this man's degrees to prove he knew all the correct answers. As the questions and answers continued one could feel his confusion, could see his complete stupidity about natural health methods. He kept saying, "Be sure to see your doctor" whenever he got in an awkward corner. He seemed to know that vitamin C helped but stated that all other vitamins were unnecessary. After stating that we should just eat a well-rounded diet and take our chances at getting all the vitamins, someone asked him about organic gardening. At this time he proceeded to say that organic fertilizers were haphazard, that the chemical fertilizers could be planned. He also stated that there was no harm in stilbesterol-fed animals if the animals were butchered correctly. This is the kind of inconsistent and stupid answers we get among the computed scientist or doctor who finds his boat rocking because of the evidence all around him. He somehow has to argue in defense of the errors he has computed because it would suddenly put him in a category of the unlearned if all this information he has spent so much time and money accumulating were suddenly to be proven incorrect. As I have stated in my other books, the physician today must use only the chemicals listed in the Physicians Desk Reference or he is not allowed to practice. The FDA controls any additions to the book, leaving us all wondering who is controlling what. It seems that if it is a chemical, no matter how harmful it may be, it somehow gets passed on to public use, but if it is plant-derived, such as Laetrile or like the herbs Doctor Meyer was using for leprosy or DMSO, it is impossible for it to get to the public market.

One of the great herbs discovered in the last decade for pain, Dimethyl Sulfoxide or DMSO, has been held back by FDA since 1960. It is made from the cement substance of trees, which puts it in the herbal class. Therefore, it is unacceptable. DMSO has been used on 50,000 patients, but that is apparently not enough. These are some of the things, discovered in tests, which Dimethyl Sulfoxide DMSO is helpful for when made from a natural substance:

Retarded children
Brain damage
Analgesia
Anti-inflammative
Muscle relaxant
Diuretic
Increases antidotes
For soft tissue injuries
Peyronies disease
Acute epididymitis
Regressive growth influence
Used in veterinary medicine
In angina prevents damage
to heart muscle
Brain and spinal injuries in
animals
Controls inflammation
Reduces pain
Burns
Arthritis
Pain of broken bones
Clears up gum conditions
Disc disease pain

Clears up inflammation and
wipes out infection in
decayed birth
Skin transplant - no scar -
fracture pain swelling

If a person drinks liquor after treatment, the DMSO increases alcohol and causes drunkeness with only two drinks. A sprained ankle can be healed in three days instead of a week making it very useful for athletic injuries.

One thousand two hundred articles on DMSO have appeared in medical journals all over the world. It was approved in Germany and Austria in 1965. It is now being synthetically made from coal tar and petroleum in those countries. We as a people have allowed laws to be passed which have boxed us right into a corner. We have allowed politicans and the bureaucracy to make laws which take away all of our freedoms, how can we progress? How can we learn anything new? How can we re-evaluate old truths to our benefit? If these laws remain and lack of respect causing uncertainty about existing medical methods continues with mounting costs, people could slip back into old superstitions as has been done many times in history. Whether or not medical science will openly acknowledge its failures with drugs, grievances toward the medical world are beginning to build to a national sentiment and will one day be more than a joke on T.V. or a movie that pans hospitals. It will culminate in a bitter contest where medical science will finally rule and people will drift off into suspension. Or natural methods will win and scientists who have held on to their computed information will stand red-faced, exposed to the truth. Before we take a look inside of the body through the iris of the eye we need to take a good look inside our country's medical practices and find out just why iridology is not accepted.

> When men like Doctor Lyman Smith of Elgin, Ill., Doctor V. Mooney of Downey, Calif., orthopedic surgeon, Doctor Joseph E. Brown who wrote in Clinical orthopedics about the enzyme chymopapain, made from papaya and Doctor Charles Hech, executive director of the American Academy of orthopedic surgeons in Chicago, all speak favorably for the enzyme for helping back disc problems and are still not able to get past the

FDA to help with a treatment made of papaya fruit and when Doctor Krebb, who made laetrile out of apricot pits has the same problem. Is there something or some-one who is trying to keep us from the wonderful dis-coveries to be found in biological medicine and hold us bound to the side-effects of drug therapy, could we be beginning on the last struggle of a weakening dictator-ship which is losing power with the people because it does not get the job done?

As we watch what has happened over the past 50 years of drug therapy in changes in the average iris as compared to the things earlier iridologists saw, we begin to understand the mag-nitude of corruption and inefficiency which has been a plague rather than a progression. As you begin to understand iridology, you will find a whole new world of wonderful but horrifying truth opening to your consciousness. You will begin to realize that you can no longer relinquish the responsibility of your body to the care of another person but you will also learn that preventive medicine and know-how is a supreme asset.

# CHAPTER TWO

# THE METHOD

In order for iridology to be of value to the lay person so that he may be able to analyze the inside of his body it would also be essential to understand how to use iridology in conjunction with herbs and correct diet in a specific way. This chapter is designed to do just that. It is not my intention to teach all facets of iridology but rather those essentials which will help each person to help himself after determining from his eyes those things which are wrong. The eyes are the windows of the body. Every error, injury or pathological disturbance which takes place in the body finds its way to the windows, where we can peer in to discover the uttermost dark, secret and hidden places. As the body heals through correct principles, the eyes will change like a kaleidoscope, a wondrous miracle to behold.

In my books *Is Any Sick Among You* and *No Side Effects*, fasting, the use of mild foods, correct food combinations, home remedies with herbs and vitamins, and care for the colon are discussed. The lay person will now learn to be specific in determining what is wrong in the body. To learn iridology, without learning what to do for what is seen, would be of little value to anyone other than to play parlor games, and such games could hurt more than they could help. As we begin to take our first look into the body, we need a good eye lupe and flashlight in order to see.

To see one's own eyes, use an eye lupe directly against the eye and a magnifying mirror in a bright light. A darker shade than the color of the eye will manifest itself as an outline on the colon, stomach, autonomic wreath and peripheral, or outside, line of the iris. Anything other than these darker lines in any way shape or form showing on the iris of the eye is disease or inherent weakness. When it is inherent weakness, there will be a darker shade than the color of the eye, either in a shadow or a bowed-out place which we call a lesion.

(1) **TEXTURE** *See Picture Page 51*

If the texture of the iris is like fine silk, the body is inherently strong; if it is loosely woven with these darker shadows showing in between, whether in a lesion or a shadow, this shows inherent weakness in the area where the darkness occurs. We can determine by this the body's strength and ability to heal. The finer the texture, the better the ability to heal.

(2) **FOCUS LINES** *See Picture Page 51*

Sometimes we may see an older person whose texture is excellent, and yet the lines of the eyes have become flat rather than sharply in focus. This indicates the body had a good inherent health but is growing old. On the other hand you may see a young person who has a sharp focus of the lines of the eyes but loose texture. This young person could have more strength than the old person even though he may be inherently weaker. His body will not last as long or remain as well as long as the older person's whose body has had fine texture. A young person can have a flat focus and a fine texture merely showing weakness. When a young person has a flat focus and a loose texture the weakness is compounded. The first thing we look for is inherent strength or texture.

(3) **THE HEART** *See Picture Page 51*

The next thing we look at is the heart. There should be no marking of any kind showing on the iris in the heart area unless there is something wrong.

As far back as 1964, cardiovascular diseases killed 921,500 people in the United States annually. When the prophets have said in the last days that men's hearts would fail them, it appears that we are living up to latter-day expectations. There are several chief types of heart conditions:

(a) Arteriosclerosis: hardening of the arteries, resulting in high blood pressure.

(b) Hypertension: causing prolonged constriction of vessels and arteries, resulting in high blood pressure.

(c) Rheumatic heart: caused by an acute inflammation resulting in damage to or having an adverse effect on muscles and valves.

(d) Congenital heart: born with diseased heart or imperfect heart, unable to perform correctly.

(e) Mitral slenosis: hardening or narrowing of valve between left atrium and left ventricle, can be congenital or caused by calcium build-up.

Hardening of the arteries is caused by scar tissue from surgery, infection, or fatty material (cholesterol) lining the artery walls, causing the opening to become smaller. This causes pressure to build, like putting a nozzle on a hose — increasing the force at which it moves in the arteries. The break occurs in the heart or the brain as in a stroke due to this force. Calcium salt deposits along the artery walls in the rough places cause a hardness and the inability of the artery to flex and perform normal peristaltic action. Hardening of the arteries causes several illnesses including heart attacks: normally smooth artery lining becomes rough and narrow because of patchy deposits which interfere with the flow of blood. The rough surface plus the sluggishness of the blood as it moves through too narrow a channel sometimes causes a clot, called Thrombosis. It may block an artery altogether. Sometimes clots break away from the spot where they form and are carried to other parts of the body in the bloodstream. The moving clot is called an Enbolus. The coronary arteries that nourish the heart muscle itself are more frequently effected by artery hardening than any other. Hardening of these arteries causes a decrease in heart muscle blood supply and may cause the chest pains known as angina pectoris. Coronary thrombosis is the term used when heart attacks occur where a clot forms and occludes or blocks a coronary artery. The aorta is the largest artery. Two coronary arteries supply heart nourishment.

A bowed-out, open or closed lesion with only a darker shade than the color of the eye may show on the heart area showing organic heart weakness. A brown spot on the lesion or just a brown spot in the area of the heart shows deterioration of the heart. Deterioration or cardic damage shows as brown or lesions in an elliptical formation. When there is a splitting separation in the autonomic wreath this shows structural or organic weakness of the heart. When there is a pathological disturbance of the aorta, the autonomic nerve wreath seems to separate immediately above the heart area. A thickening of the autonomic nerve wreath at the heart area indicates a tension or spasticity of the heart, neurogenic in origin. If there is an incursion of the autonomic wreath towards the pupil at the heart area, again disturbances in cardiac functions would be of a neurogenic origin. A distention of the colon with an incursion out into the heart area indicates that cardiac disturbances will often result because of gas pockets or impactions in the colon. This causes fibulation or irregular heart, resulting in eventual heart attacks.

Often there will be no cholesterol showing in the eye and yet the person has a heart attack. In this case it would be of a neurogenic nature, causing a spastic condition of the heart due to overwork, too much stress or nervousness. At this time the autonomic wreath would show this spasticity by an incursion toward the pupil at the heart area of the iris, 14-15 minutes in the left iris.

(4) **CHRONIC COLOR-TOXICITY** *See Picture Page 51*

After deciding on the texture, focus and heart the next in importance would be the bowel and the autonomic wreath. Death begins in the colon. If we had a clogged up sewer system in our house, we would have more sense than to continue turning on the tap or flushing the toilet, but when it comes to our human sewer system, we do not seem to have any conception of its functions. A young medical doctor writing for a local newspaper answering an old woman's inquiry regarding constipation, told her it was normal to go as long as five days to a week without a bowel movement. This is the kind of illogical reasoning which is being taught today. When looking in many eyes it will soon be discovered that most chronic deterioration begins in the bowel and along the autonomic wreath. It will then logically be concluded that the bowel must be kept clean and emptied as often as food is eaten. If you eat three meals a day, you should have three bowel movements a day. If you eat three meals and have one movement, you are becoming constipated not only in the bowels, but in the entire body waste is backing up in the lymphatic system and having an extreme effect on the brain — a reason for so many headaches and slow thought processes. You will observe that as the body becomes chronically ill, brown splotches or spots begin to appear in the iris of the eye. The iris indicates as disease progress, a definite change in the layers of the eye, color-wise; for example, the top or main color of the iris is the natural color of a given eye, blue or brown, etc. When we see inherent weakness either in a shadow or a bowed-out lesion, the color appears to deepen to a darker shade than the color of the eye. If the body begins to become chronic, the color changes first to a yellow brown, then to darker and darker brown to black, which would be complete deterioration down to the bottom layer of the iris. As the correct diet, vitamins and herbs are followed, the miracle occurs as the iris color changes back from black to brown to light brown to yellow to white to gone. The changes are simultaneously taking place in the body on a continual cleansing basis of good and bad days. When the stage of white is reached often a healing crisis will purge the remaining waste from the body and the eye will clear to its normal color.

### (5) COLON AND STOMACH *See Picture Page 52*

The stomach appears on the iris next to the pupil in the first zone. The color appears around the outside of the stomach line with the transverse colon cut in half at the naval, showing half of the transverse colon on the right eye and half on the left eye above the pupil. The ascending colon appears on the right eye para-limbal region adjacent to the stomach area on the ear side of that eye. The small intestines would be then in the center para-limbal region of the stomach area on the nose side of each eye. Wherever the colon or small intestines appear in any other shape than shown on the chart — either an incursion or an excursion with a pocket or sharp points in or out — this indicates a distortion in the colon itself. When there is an incursion toward the pupil, it indicates spasticity, a pinching off of the colon, shrunken in that area. An excursion away from the pupil out of the normal second zone indicates a lack of tone or a distention. Often in this distention there will be brown or black ballooned-out pockets deteriorating, showing the colon to be filled with all manner of toxic chronic waste accumulated possibly for years. This waste clings to the pocket walls of the colon, hardens and fills with worms. If there are pin worms involved, many black dots will appear like black pepper on the color area, if other microscopic parasites are involved, the radii solaris line will seem to project out from the inside of colon line itself. Some radii lines will seem to project from the pupil out like spokes in a wheel, these dark sharp-cut lines will often be confused with darkening lesion in the bowel or stomach. In order to change the appearance of a chronic colon on the iris of the eyes because of changes taking place in the colon itself, it is often necessary to take colonic irrigation along with the mild food diet and cleansing herbs. The herb psyllium, taken one to two tablespoons a day in water or juice, will form a soft bulk attaching itself to the hardened fecal matter in the chronic pockets, pulling off hardened rope-like putrid and rotten waste that has possibly been there for years. Psyllium will then heal the torn places, slippery elm and okra are also healing and soothing to a sore colon. Golden seal and cayenne are the herbs that stop internal bleeding; golden seal is also antibiotic. The cleansing herbs, along with psyllium, a mild food diet and some enemas, can bring the colon back to its normal functions, which is to carry off most of the body waste. Remember, death begins in the colon. Whenever the bowel is chronically ill, vitamin B6 is not produced as it should be to digest and activate our septic tank. The body would be lacking in the natural flora or friendly bacteria. Vitamin B6 added to the diet through acidophilus cultures activates and assists the large and small intestines to accomplish their purposes. Any chronic deterioration on the small intestinal area is a good indication that food is not being properly assimilated. The absorption of nutrients to be used in the body takes place through the small intestinal walls by means of osmosis into the bloodstream. If these walls are clogged with tumors, toxic waste, cholesterol or hardened oils, the absorption cannot accurately take place no matter how good the diet may be. When any chronic deterioration shows on the black spotted area shown on the chart as pey-pat (peyers patches), damage could have been done to this area when there was a high fever as a child. This type of person will always have to take more vitamins and minerals in order to maintain health or growth. Some children do not seem to grow well, and it is found to be for this cause. If the area of the small intestines is only chronically brown or black for other reasons, a cleansing of this part of the body becomes the first and most important step in order to regain health.

Iridology is probably the greatest proof of the effectiveness of a mild food and herb diet. Anytime any brown or or yellow begins on the iris, this shows that the body is deteriorating in that area.

The mild food diet and cleansing herbs are used to clean, changing the picture on the iris. Any good cleansing formula may be used or any of the cancer formulas listed in *Is Any Sick Among You*. It is my opinion that there is only one disease and that is an accumulation of toxic waste in the body. The answer is always the same: Clear out the body with herbs and mild food whenever this accumulation begins to show up on the iris of the eyes.

(6) **AUTONOMIC WREATH** *See Picture Page 52*

The autonomic wreath becomes the next in importance when viewing the iris. It is with difficulty that the body will heal when the autonomic wreath is chronic in color. The cleansing of the body becomes very important in order to move the toxic wastes from the nervous system so that other parts will clean more rapidly. If a radio had corrosion on many of the wires, even if you plugged in and turned on, the radio would not play well. The nervous system is a very intricate and vast system; its messages are carried as secretly as conversation on a telephone wire. Nerve cells look like a drop splashed on a hard surface with tentacles flung out in all directions. The cell body is the central blob, the party sprayed out resembling tree roots are called *dendrites*. Another long single fiber attached is called an *axon*; it ends in a brush-like tip. The complete unit is called a neuron. Dendrites accept incoming impulses and carry them toward the cell body. The axon carries impulses away from cell body. Nerves vary in diameter according to how many axon fibers are bundled together like cables. Each fiber is an extension of a cell body that may lie a great distance away. The sciatica nerve, for instance, runs from the back to the tip of the toes. Most axons are covered by a white fatty material called Myelin. Myelin is like the rubber insulation used on electric wires and may serve other purposes.

Billions of Neurons are found in the body — some 12 billion in the brain alone. This is fortunate, for medical science says we cannot regenerate nerve cells. A limited amount of nerve repair is possible as long as the cell body remains intact. The Neuron functions in such a way that if a nerve fiber is cut or injured the part attached to the cell body remains but the part beyond withers away. Sometimes a live part can extend itself through the withered component to reach its original destination and restore function. In some cases several nerves can be rejoined by a surgeon but only when nerve tracts are accessible; many unfortunately are not.

Electricity follows a continuous wire at a speed of 186,000 miles a second; nerve speed impulses travel only 200 miles an hour. This is fast, but not instantaneous. We cannot safely rely on quick reflexes to get us out of trouble. It takes about 1/10 of a second for nerve impulses to warn us and another split second to act upon the information, except where extreme fear is involved. In the latter a shot of adrenalin floods the system. Unfortunately our fear antenna is only geared to certain reflexes by what has been computed into our brain through or by great parental love and training (reactions to a child's cry, etc.).

Nerve impulses travel in a chain reaction, one neuron triggers the next like a train of gun powder that ignites point by point as heat reaches each neuron. This is an electrochemical process which depends on movements of molecules back and forth across membranes of a nerve fiber. The inside of a resting fiber is negatively charged, the outside positively charged. This is caused by the relative abundance of potassium ions in the interior and sodium ions in the exterior. The firing of a neuron changes the permeability of the membranes with an inward flow of sodium and an outward flow of potassium, reversing the electric charge. Almost instantly, sodium, potassium movements are reversed and so is the electric charge. These alternating movements of molecules across membranes produce currents that propagate a nerve impulse. Reversals occur as frequently as 1,000 times a second; thus, a nerve impulse is a series of tiny electrical brakes which travel over fibers and are boosted from one point to the next by chains of electro-chemical relay stations which give local reinforcement of power.

A given stimulus either causes a single neuron to discharge or it does not. Communication between individual neurons is virtually infinite in number since the cells are not physically attached to each other. Such loose connection permits uncountable numbers of switching arrangements. The junction where fine terminal branches of axon come close to dendrites of another neuron, is called synapse. There is no direct contact; a synapse is like a spark gap which a nerve impulse must jump across. This jump is helped by a body chemical, acetylchloine, which is essential for nerve impulse transmission. An excess accumulation compels various organs and tissues to run wild, causing a paralysis-like effect (like nerve gas) and a toxic build-up. We have a natural defensive enzyme called cholinesterase which breaks down acetylcholine as fast as it is formed. A fantastically rapid nerve ending build-up and breakdown of opposing chemicals keeps nerve impulses flowing in good order synapses.

When too much toxic waste is accumulated, the body must operate rapidly to keep up with this, reminding one of an old-fashioned movie being run too quickly. When this function becomes worn out and fatigued, what happens then? What happens when certain glands do not function because of the lack of certain hormones or because of overwork, causing fatigue of the glands? What happens when calcium, sodium and potassium balances becomes imbalanced? There is no difference whatever between nerve impulses except in frequencies per second. There is not a special kind of nerve impulse for seeing and another for smelling or tasting. If some incredibly mischievous rerouting of nerve pathways should occur, we would perhaps be able to hear a flavor or smell a sunset. Something of this nature arises because of glandular imbalance or growths in nervous tissue which may cause unreal but vividly perceived sensations. These are hallucinations of taste, vision and, indeed, all senses which have no detectable physical basis. This, in my opinion, is what drugs do (LSD, liquor, etc.). Could it be that drugs settle in spots when nerves cannot function, causing a re-routing or an obstruction in transmission? Could this be how acupuncture works using an inorganic needle in these places? Since the electrical system works totally on organic substances, could it be that an inorganic substance can clog, or when placed in a correct place, cease transmission?

How important, then, is the autonomic nerve wreath on the iris? In my opinion it is of the utmost importance. This area gives an exact accounting of how far the entire nervous system had deteriorated or is healing. White would show an acute condition, and is interesting to note that when there may be nothing else wrong appearing on the iris and the wreath is white, the body can be in a great deal of pain. It naturally follows, then, that cleansing herbs and mild food clean the body of toxic waste accumulated throughout the nervous system. As this is cleansing, other parts will clean also, but a complete cleansing of other affected parts of the body does not take place until the wreath is clear. In order to strengthen the nerve activity as it is being cleared, vitamin B complex and calcium of a lactate or gluconate type along with nervine type herbs, will increase the body's ability to throw off the waste.

(7) **NERVE RINGS** *See Picture Page 52*

Nerve rings appear as a partial or complete line all the way around the iris. The usual white lines indicate that such rings come and go as there may be inflammation or nerve tension in various parts of the body. A chronic brown color in the rings indicates they have been there a long time. Nerve rings can appear where there is no problem with the autonomic wreath and vice versa. Where the nerve ring crosses the spinal area it can be determined exactly where a chiropractic adjustment will be needed. When the ring follows all the way around the iris in a connected manner the entire body is affected by a nervous condition. When there are three rings all the way around, the subject is ready for a nervous breakdown. When there are four, insanity could one day ensue. There is a more serious condition when rings and wreath are both affected, and paralysis will often appear this way. There will be other things to look for, however. Markings of a chronic spot or lesion on the medulla in the brain area; markings on the spinal area — paralysis will often appear in this case as a thickening of the autonomic wreath, giving the colon and stomach area a three-dimensional appearance or of being caved-in. There will also be markings showing on legs or arm areas if there is a paralysis in either.

It is extremely important to maintain cleanliness of the body tissues in paralysis, especially the bowel as elimination is often slowed down to the point where the body deteriorates very rapidly if this essential process is not taken care of. It is also important to feed the nervous system those elements which soothe and strengthen. Nervine relaxant herbs such as red clover, lobelia, hops, catnip, valerian, chamomile, etc., and the vitamins and minerals necessary to feed the nervous system such as B complex and calcium lactate or calcium gluconate. Of course it is important to remove the cause of stress, if possible. Often a person will not even realize that certain things in his daily life are irritating his nervous system.

(8) **GLANDS** *See Picture Page 52*

The various glands of the body are in trouble when any

marking or lesion appears on the iris of the eyes. The lines of the eyes are normally straight from the autonomic wreath out to the peripheral line of the iris. Wavy lines indicate glandular imbalance. We then look to see which glands are involved. If there are no markings or lesions showing on any of the glands but the lines are wavy, it is always adrenal imbalance caused from too much stress, a traumatic shock, or a great or extended unsolvable stress.

When the herbs related to various glands are used in a correct balance, the lines will straighten. If there is deterioration in any glands, it is necessary to clean the entire body with mild food and cleansing herb formulas along with specific hormone herbs. Hormone herbs are as follows:

Pituitary:
Alfalfa — Approximately 10 to 20 or more tables or capsules.
Adrenal:
Licorice (Acts like natural cortisone) — Approximately 2 to 4 capsules, with mild food. On regular diet and meat, up to 15 capsules.
To get off of cortisone and onto licorice: — Approximately 15 capsules a day until withdrawal is past — approximately 10 days.
Female:
Black Cohosh (Acts like natural estrogen) — Approximately 1 to 3 capsules; if headache, cut back, it is not needed. If a nervous condition persist, another gland is involved.
No problem or withdrawal to get off of estrogen and onto black cohosh.
Male:
Ginseng — Approximately 4 to 6 per day no real limit
Sarsaparilla

Pancreas:
Golden seal, Juniper (or cedar berries) (Diabetes) — Approximately 1 to 30 capsules. On a mild food diet, only small amount is necessary, approximately 2 to 4. Juniper berries: Approximately 6 to 8, when on mild food.
Hypoglycemia:
Licorice
Hyperglycemia:
Insulin overshoots mark, drops to low blood sugar, high and

low periods daily or hourly.

Will be seen on iris with adrenal as well as pancreas markings. Other glands may be involved, also can eventually become diabetes and Addisons disease.

Licorice, Golden seal — One-half as much as the amount of licorice used.

Adrenals:

When the adrenals are involved, often the thyroid and pituitary and female area also show markings. Eventually becomes Addisons disease. Golden seal usually not necessary unless sugar is used, causing a pounding heart — steady, consistent low fatigue.

Thyroid:

Will show a marking on the thyroid area, lesion or deterioration — check for fluid build-up and uric acid, acid stomach and related adrenals.

Kelp, dulse — Hyper, thin type, approximately 4 to 6 kelp tablets.

Kelp or Dulse liquid — Hypo, fat type, approximately 20 drops up to 70 or more, look for fluid build-up on iris field. Potassium gluconate or diuretic herbs to regulate fluids and help thyroid gland in its function.

No problem to go off of thyroid drugs onto herbs, no withdrawal; higher doses necessary to regulate, approximately 20 drops and more, daily.

Dr. E. Hugh Tuckey gave us the valuable information of hydrochloric acid (Footnote as found in National Health Federation Bulletin October 1967) for use in digestion. He said that the major role of hydrochloric acid is to digest protein and if there is too little hydrochloric acid, the stomach does not empty as soon as usual and foods eaten 3 to 4 days earlier, not digested causes putrifaction, causing gas, a belching condition and a burning in the stomach, a substernal burning just behind the breastbone. He says further that a person with these problems will then take a bicarbonate of soda or other type of antiacid to alkalize a stomach that already does not have enough acid, when often the anti-acid used will be an aluminum preparation which he says can cause cancer.

He states also that on tests, lactic acid shows up in the stomach in these cases and it does not belong there at all. When the emptying time of the stomach is slowed down due to hypo- or anachlorhydria, he says we always find a spastic pylorus and we can demonstrate this clinically one hundred times out of a hundred; we don't have to guess at it. Then he has shown how this spastisity in turn involves the bile duct and pancreatic duct. If the pylorus is in a state of spasm, the duodenum is also in a state of spasm so a normal amount of bile is not emptying into the small intestines. One of the properties of bile is a bacteriacide along with promoting the peristalsis of the small intestines. This also causes the liver to become engorged with bile that should be going to the small intestines. The pancreas is also affected, causing a backing up of its secreting substances and the pancreas then becomes enlarged, hard and extremely tender. We can see that this lack of hydrochloric acid has far reaching effects involving the stomach, liver, gall bladder, pancreas, assimilation of food, constipation and pancreatitis and even hepatitis. Doctor Tuckey gave hydrochloric acid to his patients so that people could handle the protein and found that they would have to take these on an indefinite basis unless they were given a pancreatic, duodernal substance, then the hydrochloric acid would normalize in two to four months. Doctor Tuckey was not clear on the cause of the problem but rather thought it could be caused from emotions and the stress of every day incompatible living conditions. He has given us some very valuable information resulting from thirty years of research. He mentions also the use of hydrocholric acid tablets when traveling in countries where there is bad water, for killing the amoebas in the water and preventing diarrhea discomfort while traveling. It is with appreciation that I mention his findings and I wish to add rather than distract from his work. It has been by discovery that the cause is as he had supposed an adrenal cortial exhaustion causing a lack of the adrenal cortine hormone which interferes with the production of hydrochloric acid, causing a uric acid build-up in the stomach. This is also the cause of the lactic acid build he mentioned resulting from any exercise. The lactic acid builds in the muscles from exercise and the excess normally routed back to the liver and is changed to glycogen and used again as energy. This seepage through the abdominal wall could result because the liver is too badly clogged in the way he has described and so seeps into the abdominal tissue and into the stomach, causing the burning sensations experienced when lactic acid is present in the stomach.

The experience I have had shows me that licorice root acts like the adrenal hormone relieving the mental stress but this hydrochloric problem is relieved by saffron, to open the clogged liver, having a definite effect on oil, and assist the pancreatic functions and then dandelion dispells the uric acid which has built up in the stomach and the rest of the body — somehow regulating the hydrochloric as a natural magnesium or anti-acid effect on the body, without the side effects of taking an anti-acid such as soda.

(9) **OPEN LESIONS** *See Picture Page 53*

Open lesions are inherent weaknesses when they are just a color shade darker than the color of the iris. Any darkening shades down into lower layers indicate chronic deterioration in varying stages. Open lesions heal faster than closed lesions. Lesions that are chronic will sometimes knit, having the appearance of a darned sock. Lines will weave closer and closer at times closing the lesion when corrective measures in diet and herbs are taken; other times they may clean to a color shade darker than the eye and remain. Chronic deterioration begins usually on the outside edge of the lesion moving from outside to center in varying stages of chronic color from yellow brown to black but as it begins to heal, white will appear on the top and around edges with the knitting appearance. Colors underneath gradually lighten and colors on top become whiter and whiter until they disappear.

(10) **CLOSED LESIONS** *See Picture Page 53*

Closed lesion is more serious than an open lesion and is more difficult to heal. When white knitting lines begin to appear it is a good indication that the problem can be solved, otherwise the same would apply to the closed lesion as with the open lesion.

(11) **PARASITES** *See Picture Page 53*

Parasites are often the reason people who have fasted off and on for years and lived on a good diet never quite reach the stage of good health they work so hard for. Parasites live on the refuse of the body and hold onto the waste even in a fasting state — especially in a fasting state, as it is their food supply. The only time they turn loose of it is when they are killed; then the mucus in which they live sloughs off like a healing crisis.

The radii solaris spokes which look like a dark, sharp cut line on the iris indicate heavy parasite involvement in the particular part of the body where they are showing. When the parasites are killed the lines begin to flatten and fuzz off; then disappear altogether. The so-called radiation lines which are wider than radii solaris spokes are also involved with parasites of a stronger variety. Pumpkin seed and herbal pumpkin formulas will usually take care of radii solaris spokes but it takes black walnut to change the radiation lines. The time it takes to get rid of them varies depending on how deeply entrenched they are. A crisis can occur when mucus sloughs off and lasts anytime from two weeks to six or eight months. Sometimes no real crisis is necessary for the mucus just begins to move a little at a time. As parasites are killed a smelly drainage is caused if it is near a body opening such as in the head area, kidney, vagina or rectum.

Parasites were a definite pathalogical concept in ancient Egyptian medicine as shown by the graphic reproduction of animals in mural paintings, comprising an important part of certain hieroglyphics that indicate disease. In Egyptian pathology the animal was considered the symbol of disease. Ra, the sun God, falls sick because of a worm born from his sputum, when the parasite is not visible, an invisible worm is imagined.

During the period when Egypt passed under domination of Persia, the decay in medicine began. The conquerors tried to preserve ancient traditions in medicine but the physician was so closely attached to the religious priests and government rule, as this changed hands the brilliant medical practices decayed into sorcery and charlatan drug vendors, similar to what has happened in the past 50 years to our society. Our grandmothers were well acquainted with de-worming their children, but today we only de-worm our dogs.

### (12) INJURY

Injury will show up in many shapes and forms with all lines irregularly crossing natural lines. Toxic waste and deterioration follows the regular line pattern of the iris from pupil to peripheral line but an injury can cross in all or any direction, this is what makes it easily discernable.

### (13) SURGERY

A surgery will look at times as though a hole was cut out of the iris in all shapes and sizes. A broken bone will usually show a straight cut line crossing the bone area where the break occurred. An injury or a surgery may, if permanent damage is done, turn a brown color and remain that way even though a cleansing diet and herbs are extensively used.

(14) **DRUGS** *See Picture Page 53*

Drugs can be easily identified on the iris of the eye by a sodium ring which is a grey-white or brown ring around the peripheral line on the 6th and 7th zones. This ring appears to be super-imposed above the top layer. Drugs are also revealed as a red or bright brown circle around the outside edge of the pupil.

What is meant by the word "drug"? In centuries past the word "drug" was used synonymously with herbs or medicine. As herbs became obsolete to the medical world and were replaced by chemical drugs, herbs were no longer classed as drugs. Some drugs of a chemical nature are made by mixing chemicals with herbs. In this case the herb is still classed as a drug because of the chemical additive. The chemicals used to make drugs are inorganic substances, such as potash, salts or petroleum. We then have another drug factor used which is made from organic narcotic herbs, such as morphine, caffeine, nicotine, etc. Some are used in the natural state, many others are used in conjunction again with chemical salts or potash. In order to identify those botanicals classed as drugs, we find that all must be named with a name ending in "-ine" whether in a natural state or as a mixture with other salt chemicals.

Hard narcotic drugs show up on the pupil red ring and whenever brown spots appear to be super-imposed over the top layer of the iris. Furthermore, drugs of a narcotic or a salt nature both have an effect on the autonomic wreath and when it appears to be chronic in color. Drugs are inhibiting the normal routing of the nerve signals throughout the area of the body where such chronic color appears on the wreath. Sometimes the entire wreath will appear chronic in color: this is another reason for a cleanse to rid the body of the inorganic or drug substances interfering with nerve impulses. This is also the reason for the side effects of drugs. When the nervous system is mal-functioning it is with difficulty that the body heals. As drastic changes in the iris occur because of the chemical salts and drugs, the dangers of drugs become more apparent. All of these markings of a brown chronic nature whether drug caused or incorrect eating habits and poor nutrition, are indicative of deteriorating disease. It makes no difference what the cause, when we see these shadows of doom registering the extent of deterioration in the body, it is time to find out what we have been doing wrong and make a change for the better.

(15) **ACUTE CONDITION** *See Picture Page 53*

An acute condition in the body is shown on the iris as white over the top of the main color layer. A condition such as a cold, flu, measles, mumps, chickenpox, etc., will all show white over most of the iris field. There are varying degrees of white: the more heavily the body is eliminating, the whiter the eyes will become, as the elimination passes out of the body, the white will gradually disappear. Some people have white on the eyes all the time. This type of person has a good ability to eliminate and is in a constant state of elimination. They are always fatigued and weak, the way one feels when he has a cold without the fever. This type of individual would be well rapidly and remain well if he would stop the intake of concentrated or poor foods which the body is always trying to throw off.

Often when the body is chronically sick and an attempt is made to fast or semi-fast by mild food diet and herbs, the body will begin to eliminate and the eyes will whiten as the elimination begins to take place. "Acute" means toxic waste on the move in the bloodstream on its way out. White always means healing whether nature is forcing an elimination as with a cold to save life or whether an elimination is caused by a semi-fast, etc.

In case of an inflammation, white would show in the inflamed area as white on the top of the top color layer. White appearing deeply in between the lines indicates parasites, coupled with an acute pain such as would be caused in Lupus, rheumatoid arthritis, or malaria. The white line may turn dark and become a radius solaris showing only that the parasites are more deeply entrenched and of a chronic nature. Because white is always of an acute nature, pain is always involved where white is showing. In the case where parasites are involved, the body can remain in an acute, painful condition as the toxic waste just stirs around in the bloodstream and is unable to leave the body. In these cases the parasites must be killed in order to eliminate the waste. As healing begins to occur white lines will appear around the spot or cris-crossing white lines where chronic disease is involved. Sometimes healing looks like someone placed a white screen across a brown spot leaving the brown showing underneath.

(16) **LIVER, GALL BLADDER** *See Picture Page 53*

Any chronic deterioration showing on the liver and gall bladder causes a difficulty with healing. Sometimes the bowel is blocked causing the liver to malfunction. The bowel area must be checked to see if it shows a toxic condition. If so, this would be the first thing to clear out in order for the liver to empty and clean. When chronic color shows in the gall bladder, gall stones could be the problem. The gall bladder purge as related in *Is Any Sick Among You* could be used to empty the gall bladder of stones so that normal elimination can be carried on. The other place to look in conjunction with the liver and gall bladder would be the duodenum. If there is an ulceration showing or a chronic color, slippery elm is the herb soothing to ulceration.

(17) **KIDNEYS** *See Picture Page 54*

The kidneys are a very important part of the body's ability to eliminate. We have four main organs of elimination: bowel, kidneys, lungs and skin. When any one of these is clogged or malfunctioning, it places an added burden on the others. It is therefore important to keep each one of these organs functioning well. If there is a chronic condition appearing on the iris either in a lesion or a spot, special care must be taken to make sure the bowel is free and clean. If there is an acute condition — white showing on the kidney area — the bowel must be kept almost loose in order that damage will not be done to the body. When a person has a fairly strong eliminative ability at the onset of an acute kidney infection, the bowels will automatically become loose in order to save the kidneys and re-route the fluids through the bowel. Lemons and vitamin C help to dissolve the mucous and any stones. Golden seal is an antibiotic for any germs, while kidney herbs listed on the chart are for restoration. When the inflammation appears a gray metallic color, an over acidic condition exists. Dandelion and saffron, chlorophyll and vitamin C dispel acids.

(18) **BONES** *See Picture Page 54*

On the iris chart the bony areas are well defined: shoulder, scapula, thoracic (rib cage), spinal, leg (thigh, knee and foot), hand and arm, pelvis, and the neck are represented (the latter on the medulla). The top of the head appears at twelve o'clock; back of the head towards one o'clock; jaw, nose, face and skin area close to the periphery. A break will cross the normal lines of the iris and can show if the bone has been shattered or a clean break. When deterioration has set in, the break will show as brown or black — otherwise cris-cross, jagged, odd-shaped markings depending upon the break itself.

Often in an old break, even one that was well set, arthritis will set in and will also appear to be brown in color. If it is a damaged deterioration of bone and surrounding tissue, the brown color may never change. If it is an arthritic condition, irreversible changes may occur in bone shape but the brown showing deterioration may leave the iris. The brown or black — if it is arthritis in the bone — will show on the sixth zone on the bony area.

Hypercalcemia will appear as a gray-white color. If it is general in the whole body, the entire iris color will appear to be gray-white. When it is specifically causing sight problems, a gray grandular appearance will appear on either side of the pupil. A calcium problem in the body, whether it is settling in the arteries, in the joints or in the tissues, involves the adrenal and parathyroid glands.

In order to regulate the correct calcium metabolism, glands need to be properly functioning and secreting their necessary hormones. When they are unable to perform accurately, either because of too much stress or organic weakness, many things begin to malfunction. The bowel begins to ulcerate, and the kidneys are then overworked and malfunction, causing fluid to build in the tissues. The skin backs up trying to carry off waste and causes a splotchy pigmentation. The liver does not function as it should, breaks or lesion begin to occur in the inner abdominal wall, allowing a seepage of lactic acid into the stomach and interfering with digestion. Lactic acid is not changed to glycogen by the liver so as to be used again as energy, and a toxic condition begins to increase. The result is complete fatigue, exhaustion, mental problems and nervousness.

Fluid build-up in the body and around heart and eyes causes sight problems and heart weakness with a possible uric acid build-up or burning sensation. It can be determined just how far hypercalcemia has gone by checking each of these organs mentioned. In order to bring the calcium back to normal ossification when the body is not producing adequate hormones, the herbal hormone-type herbs will sustain life the way insulin does in a diabetic. Where there is an extreme uric acid and lactic acid build-up, saffron and dandelion are helpful herbs. The adrenal hormone-type herb is licorice; the parathyroid hormone-type herbs are alfalfa and kelp.

Osteoporosis is a calcium lack where insufficient calcium is taken into the body and the body begins (activated by the parathyroid glands) to leach calcium from the bones in order to maintain correct calcium levels in the blood. The bones then become hollow and brittle. This often happens with old people who live on bland diets not rich in calcium and other minerals. This is a cause of much of the senility in old age, because in order for the brain to function, blood calcium must be kept at correct levels. Even in a hypercalcemia condition where calcium is deposited in the body, the person could have too low a level of calcium in the blood. This is the cause of the mental breakdown common to the hypo- and hyper-glycemia type. Whenever an arc or crescent appears across the upper portion of the iris brain area — looking as if the white were pulled down over the upper iris obscuring the peripheral line — it is an indication of senility beginning because of the lack of calcium to the brain. By helping the glands and the calcium metabolism in the body senility can often be thwarted. Sometimes, however, the arc is caused by brain damage, and senility is the result anyway. An osteoporosis or hollowing out of the bone condition appears on the body structures of the body in the iris as a pit-like appearance, like cement that has not been troweled enough. The addition of calcium and related mineral which cause calcium to be utilized in the body will readily regulate this condition.

### (19) **EYES**

The eye section of the iris would show no markings unless there were inherent weakness shadow lines or lesions. When a dark discoloration begins to appear, deterioration is beginning to occur to eyesight. Related organs that can cause sight problems need to be checked: kidneys, glands, bowels and nerve wreath. In order to help the sight, all related organs must be helped also. An infection in the eyes or hay fever, etc., will show a white line on top of the top color layer.

(20) **EARS** *See Picture Page 54*

With the ear it would be the same as with the eyes. Sometimes, however, parasites may have invaded the glands around the ear and neck to the extent that a radiation line will appear on the iris. This radiation line — or parasites — will cause eventual prostration of the nerves and can cause nerve deafness. Watch for radii solaris lines and radiation on eye area also.

(21) **ANEMIA** *See Picture Page 54*

Anemia is a word not correctly understood. Unless there has been hemorrhaging, the need for iron is a very small amount. This arc is often accompanied with a scurf rim, placing an added burded on the internal skin and other organs of elimination. This should indicate that toxic waste is accumulating in the area of the arc, and has difficulty in moving itself out. Anemia means an encumbrance of toxic waste in the blood. It is interesting to note that when the white blood count is high and the system is cleansed with herbs and mild food, the white count goes down and the red count goes up. When the system is cleaned, the arc clears off the iris.

(22) **SCURF RIM** *See Picture Page 54*

The scurf rim suggests the condition of the skin. Anytime the peripheral line is wider than a pencil line this is called a scurf rim, and it may widen in as far as the fifth zone. The uniformity of depths can vary all around the circumference of the iris, indicating that the skin may be better in one place than another. The darker and more defiled the color and the wider the ring, the greater the impairment of skin function and the retention of toxic waste. If the color is only a shade darker than the color of the iris and a scurf rim is showing to any extent, this is only inherent weakness. Other problems will result if the bowel and kidneys cannot carry off the waste properly. The skin is the third kidney, and when it is not functioning well, the entire body suffers. Wherever there is an incursion of the scurf rim, the area of incursion becomes more readily toxic in that part of the body. When the lymphatic rosary nodule appears to have spilled over into the scurf rim showing a white line, this indicates swelling or fluid build up in the tissues of that area of the body, revealing these white lines on the iris.

(23) **LYMPHATIC ROSARY** *See Picture Page 55*

The lymphatic rosary on the iris often looks like globs or nodules of white, yellow or brown beads strung around the sixth zone. Sometimes they connect all the way around. Other times they are disconnected and show only sparsely around the circumference on the sixth zone. Anytime these globs show on the iris anywhere on this zone, it is an indication of an accumulation of mucus waste in the lymph of the body. If the color is extremely white — almost to blue — and appears to be super-imposed on the top of the fibers of the iris, it is considered as an index of arsenic retention in the lymph system. Whether the lymphatic rosary appears white, yellow or brown and to have a thickening of the fibers themselves, this would indicate lymph congestion only. Often there is arsenic retention as well as mucous congestion of the lymph, making it difficult to distinguish. The arsenic retained in the lymph is related to neuritis conditions and will become neuritis later as it becomes more chronic. This is also related to arthritis.

There are seven zones from pupil to peripheral line. The sixth zone where the lymphatic rosary is found has also to do with circulation. When circulation is poor, the lymph nodes and the body begin to retain waste. As waste is retained in the lymph, the thyroid works overtime to send iodine into the body to kill the germs living in the mucous waste accumulating. When thyroid activity is increased, pituitary activity is decreased, as they act like a balance scale: increase of one decreases the other. When thyroid activity is increased, oxidation increases to assist the body in expelling waste the same way we breathe faster when we have a cold or fever. The spleen is also overworked since it acts like the vacuum system of the body, and it is then overworked causing a breakdown in its ability to do its job accurately — a lowering of resistance to infection. This is not a valid assumption, however, because all that has to be done is clean the body of toxic waste, and normal function will be resumed.

There are a few other places necessary to know not listed on the chart which occur in rarer cases. Wilsons disease causes a copper-colored ring to appear on the seventh zone or skin area. Multiple sclerosis shows markings of deterioration on the medulla and spine. The autonomic wreath and colon will appear very toxic. When paralysis has begun, markings on the arm and leg areas appears on the iris and the autonomic wreath will usually separate at the medulla area. If calcification has begun the iris color changes to a gray-white and the autonomic wreath appears to be raised, while the colon appears to be sunken in. Calcification can change the pressure of cerebrospinal fluid and have an effect on the choroid plexus which regulates these fluids. Other times gland malfunctions can have an effect on this fluid, which affects the spine and brain.

If the pressure of the cerebrospinal fluid is not kept stable, the entire body can be affected. Many pathological and functional disturbances then begin to occur: digestive, respiratory, headache, circulatory ailments causing general slowing-down of eliminative processes which result in a general toxic condition. When this happens, legs can be affected — anything from varicose veins to paralysis of legs and feet and could even begin with a gouty condition. When other glands are not functioning properly (especially the adrenal glands) the cerebrospinal fluid pressure can be lowered, resulting in the above pathological disturbances. In either case, calcification or cerebrospinal lowering or pressure are the result of glandular imbalance which will also appear as wavy lines on the iris.

When individual or collective gland problems are solved by hormone herbs and corrective diet, the calcification is halted and normal ossification begins again. Since the choroid plexus is a gland also secreting hormones, alfalfa is the regulatory natural hormone herb for the choroid plexus. An acid stomach will appear as a gray color on stomach area, but here again normal ossification is beginning to be disturbed, it is caused by certain gland imbalances of the body — deterioration in the medulla or other glands or merely an intolerable stress which throws the glands out of balance.

The medulla may be observed for many malfunctions: did the glands affect the medulla or did the medulla cause the glandular imbalance? Parasites should be watched for in this medulla or any other deterioration or inflammation. The medulla can have far-reaching effects on the body since it regulates saliva, controls respiration, controls cerebrospinal fluids, swallowing, heart, vomiting, sneezing, and sweating. Whenever deterioration shows on the medulla, any of the above symptoms or pathological problems can be the result, making it doubly important to use cleansing herbs. Certain asthma cases and even multiple sclerosis could be caused by a defect in the medulla. Twitching of the muscles is usually a lack of magnesium, but where magnesium does not stop the twitching eye or muscles, it is then caused by the lack of the parathyroid hormones calcitonin or parathormone. Here again alfalfa is the answer, regulating the parathyroids in their function. As with other brown or deterioration markings showing on the iris, cleansing herbs and regulatory hormones along with mild food are the solutions to the problem.

Let me say again there is only one disease and that is an accumulation of toxic waste. This accumulation happens because of incorrect eating habits, improper nutrition or inherent weakness. When there is inherent organic weakness, a better nutrition and more care must be taken to live a better, healthier life. When the body is filled to the limit with toxic waste mucus, the only way to better health is to clean it out. The only way to clean it out is to stop the intake of concentrated, mucous-forming foods so that the mucous waste can eliminate from the body. Herbs are a great help in this elimination process, causing the mucous to dissolve and eliminate more rapidly. Some herbs are meant to kill the bacteria and parasites living on the mucous so that the mucous they hold onto and live in to survive can be released and expelled. Many herbs can be taken together, working like a fine team — a precision machine to cleanse, soothe and heal, each one having an almost innate intelligence which directs it to the special place in the body where it functions. The greatest secret to good health is to keep the body clean inside as well as out.

Many important and knowledgeable people are coming to the conclusions that we have far too long heralded the virtues of meat protein even to the exclusion of the use of necessary mineral foods.

In an article from *Today's Health*, Daniel Grotta Kurska points out that scientists have long classified meat as first class protein and vegetables as second class, thereby implying that non-animal protein is somehow inferior. Current evidence, however, points to other conclusions.

The steps to the top of the ladder of disease are as follows, as we start up the tree of disease:

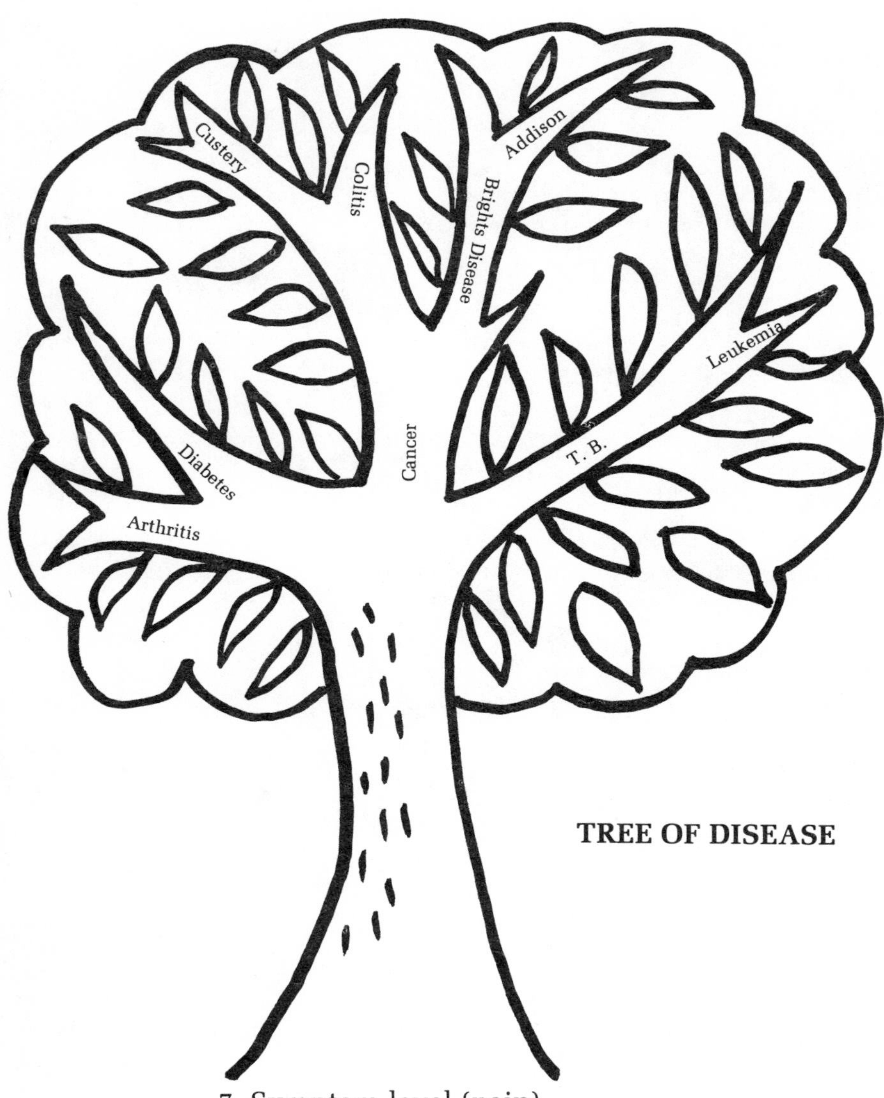


**TREE OF DISEASE**

7. Symptom level (pain)
6. Parasite and germ infiltration
5. Increased oxidation
4. Decreased pituitary activity
3. Increased thyroid activity
2. Lymphatic retention
1. Normal body chemical balance

# PICTORAL SECTION

# PICTORAL SECTION

The following pages of eye photographs have tried to illustrate the various problems found in selected iris. Pictures have been enlarged making it easier for the naked eye to detect the problems. The pictures from pages 51 to 55 relate to Chapter 2 of this book. The pictures on the remaining pages are our "Before and After" group. These are the iris of some brave souls that have discovered their problems and are striving to correct them.

Reference is made here to pictures in the "Before and After" group that have corresponding testimonies in the body of the book:

Gary Gillum — testimony page 165; bottom 2 pictures page 58

Wyonna Kruse — testimony page 175; top 4 pictures page 58

Jeannine McKenzie — testimony page 169; 3 pictures page 59 left section

Ruth Ruhland — testimony page 85; 2 pictures page 61 right section

Michael Tracy — testimony page 169; top 2 pictures page 57

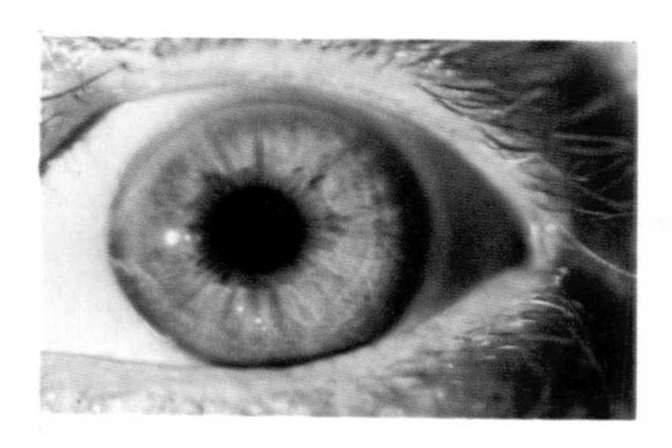

NO. 1 TEXTURE
See Page 24

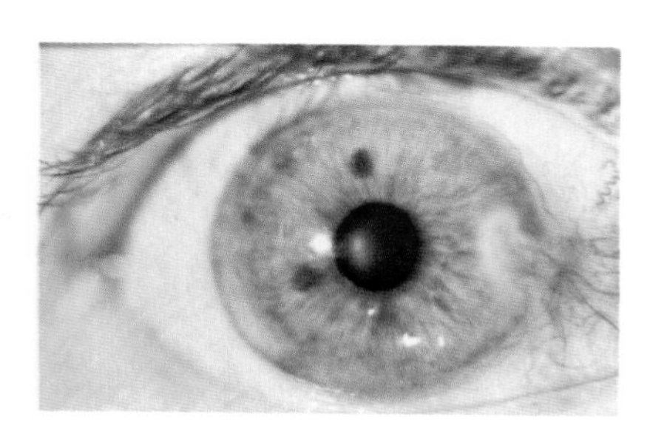

NO. 2 FOCUS LINES
See Page 24

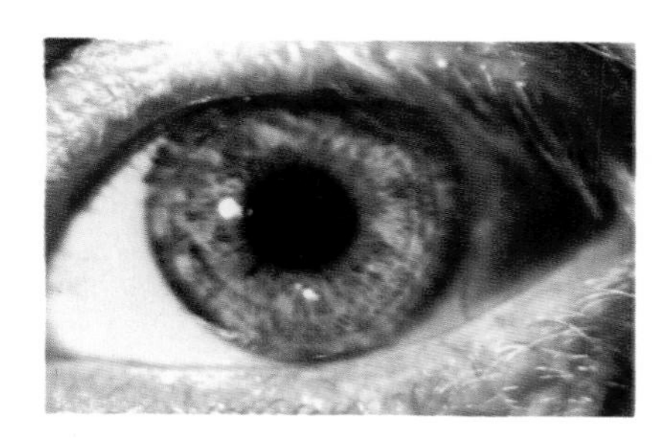

NO. 3A HEART
See Page 24

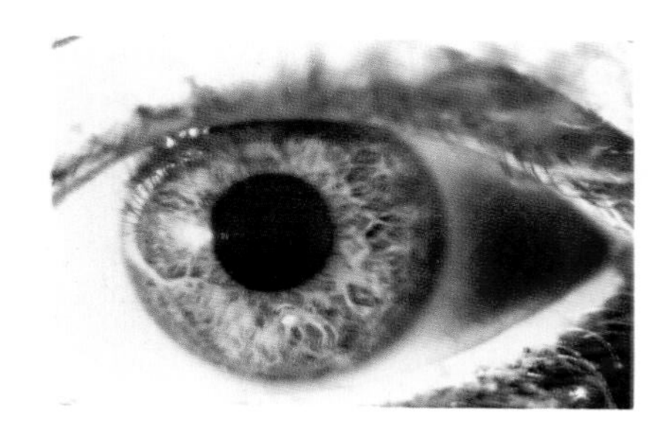

NO. 3B HEART
See Page 24

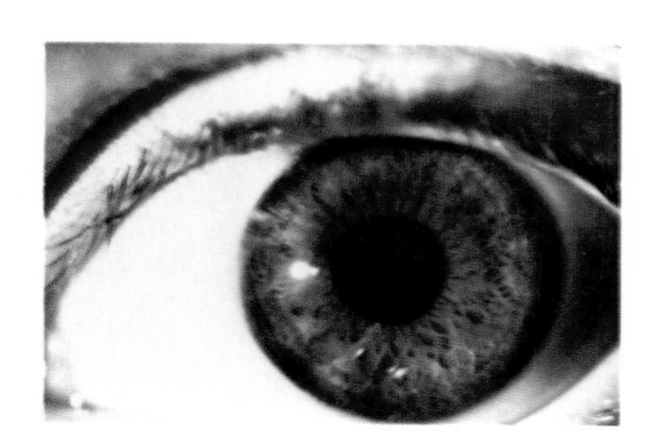

NO. 4A CHRONIC COLOR-TOXICITY
See Page 26

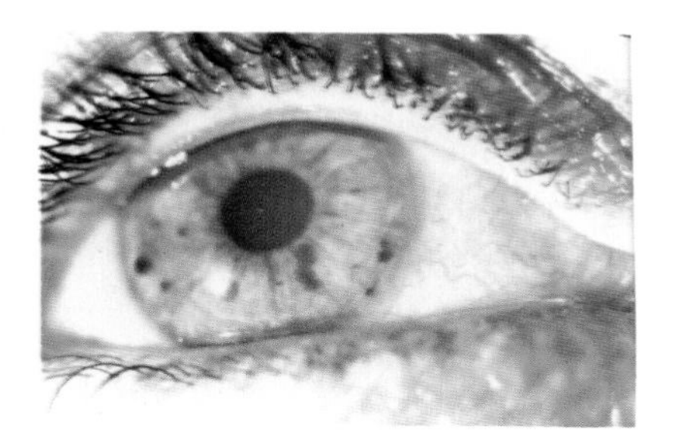

NO. 4B CHRONIC COLOR-TOXICITY
See Page 26

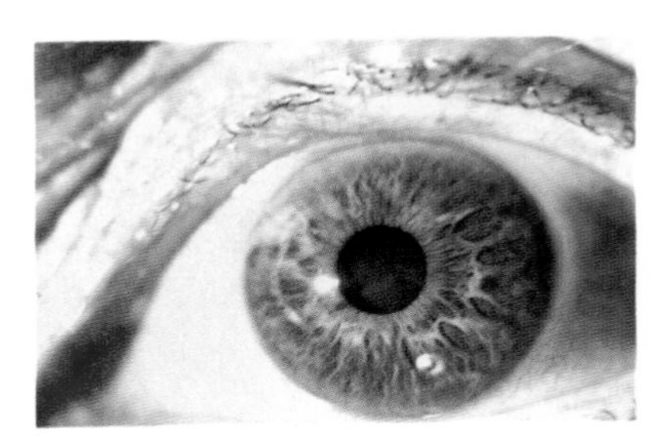

NO. 5ACOLON & STOMACH
See Page 27

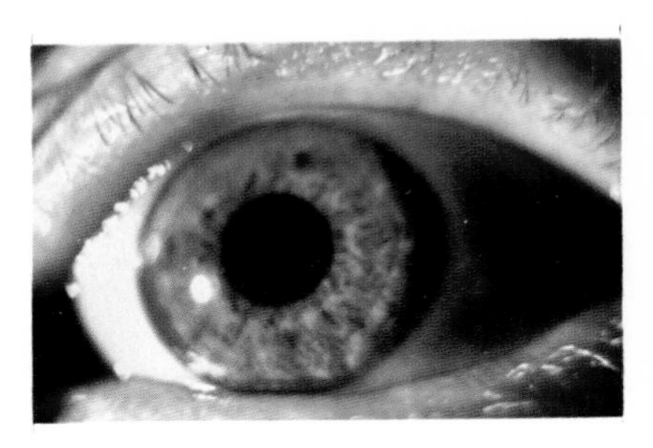

NO. 5B COLON & STOMACH
See Page 27

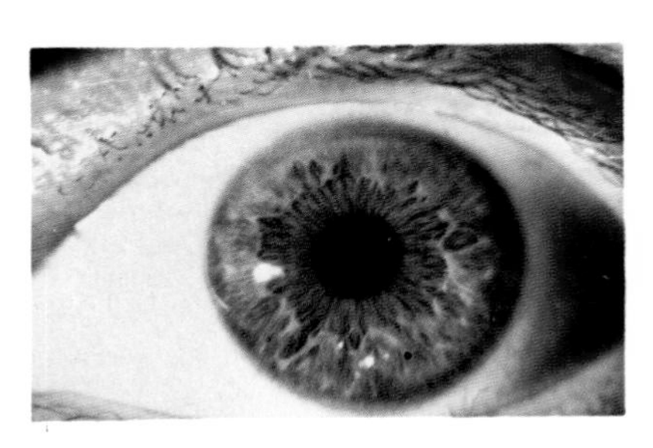

NO. 6 AUTONOMIC WREATH
See Page 29

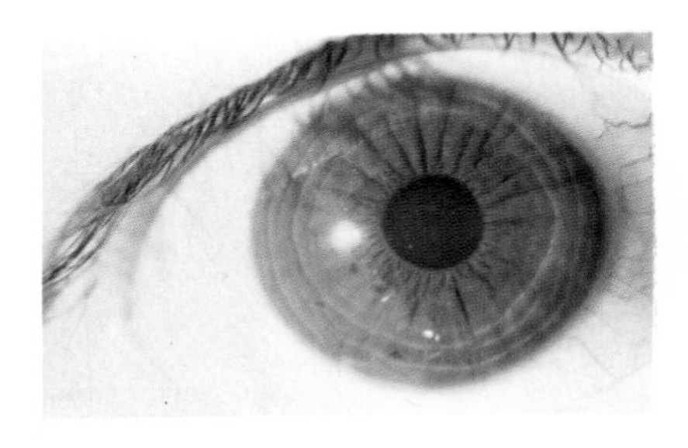

NO. 7 NERVE RINGS
See Page 32

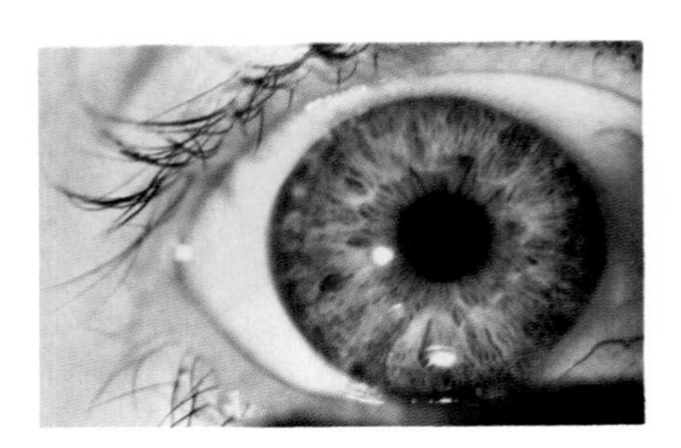

NO. 8 GLANDS
See Page 32

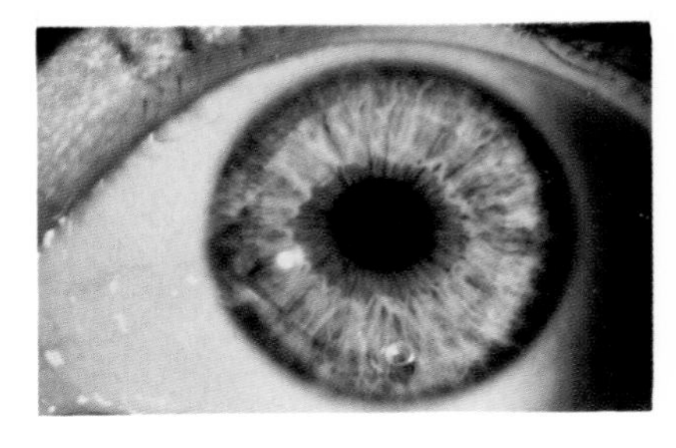

NO. 9 OPEN LESIONS
See Page 36

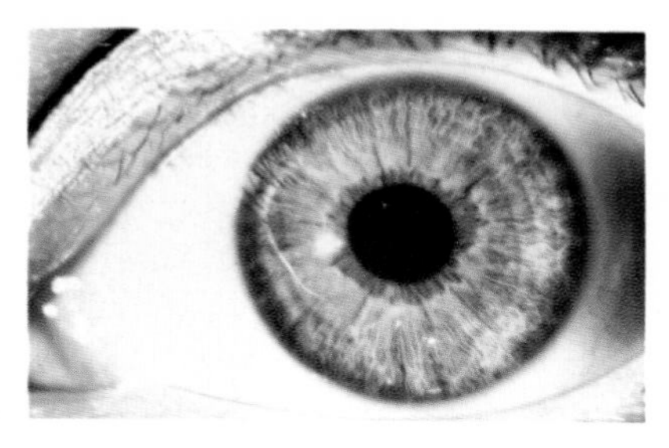

NO. 9 OPEN LESIONS
See Page 36

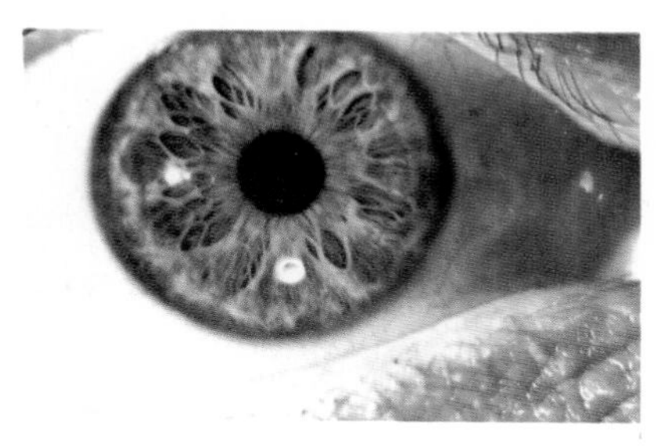

NO. 10 CLOSED LESIONS
See Page 36

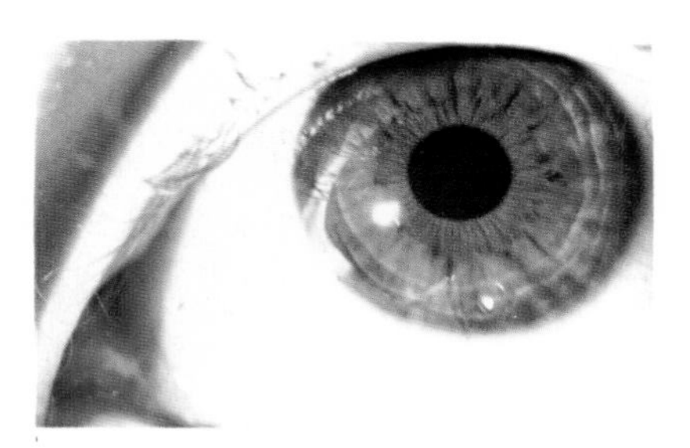

NO. 11 PARASITES
See Page 36

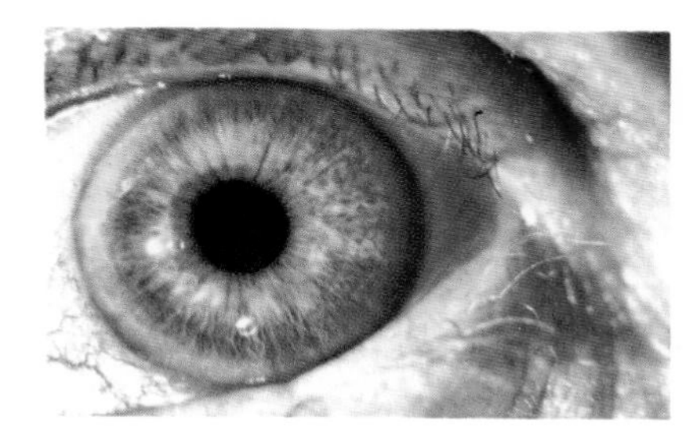

NO. 14 DRUGS
See Page 38

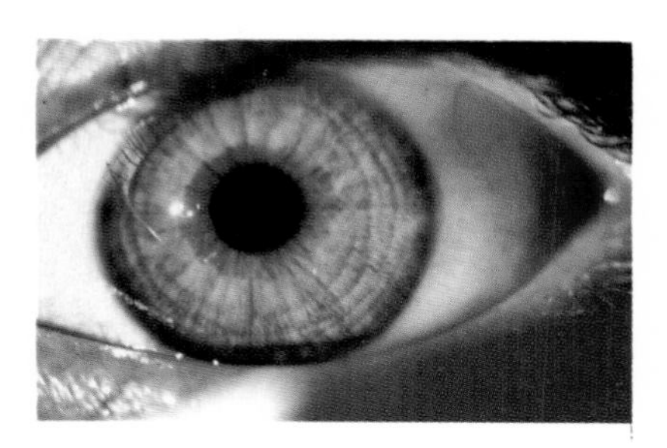

NO. 15 ACUTE CONDITION
See Page 39

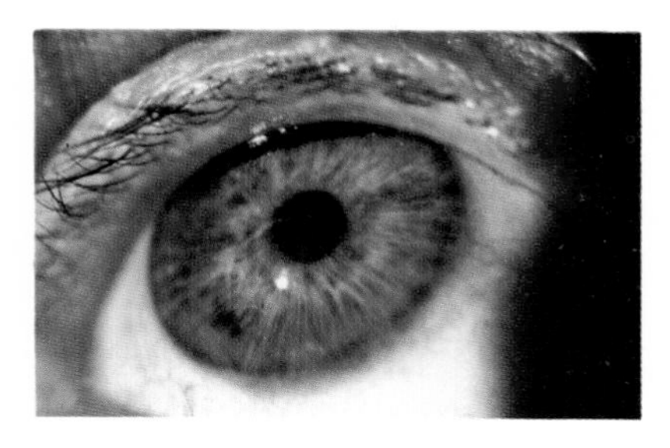

NO. 16 LIVER-GALLBLADDER
See Page 40

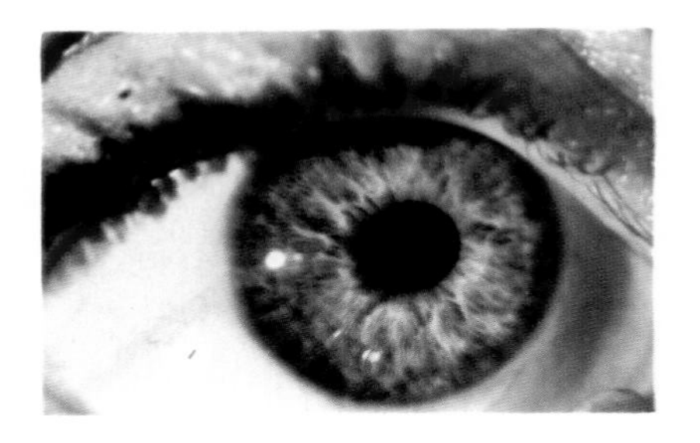

*NO. 17 KIDNEYS*
*See Page 40*

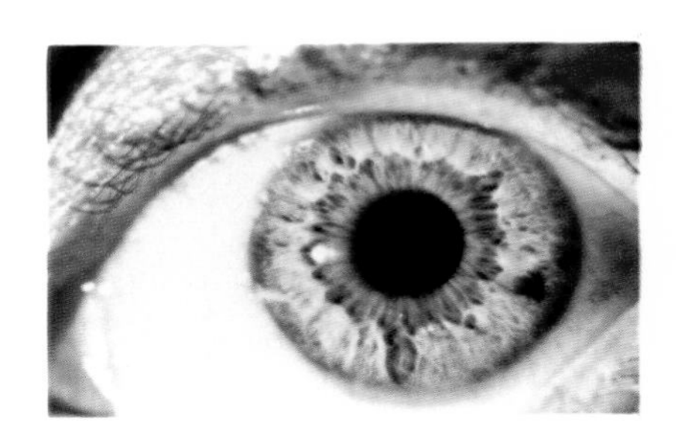

*NO. 18 BONES*
*See Page 40*

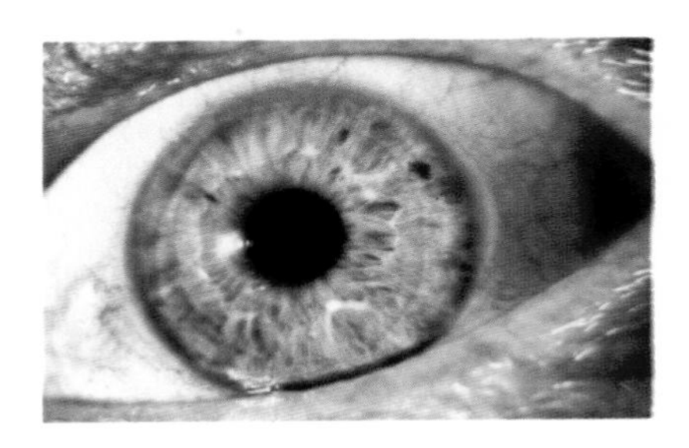

*NO. 20 EARS*
*See Page 43*

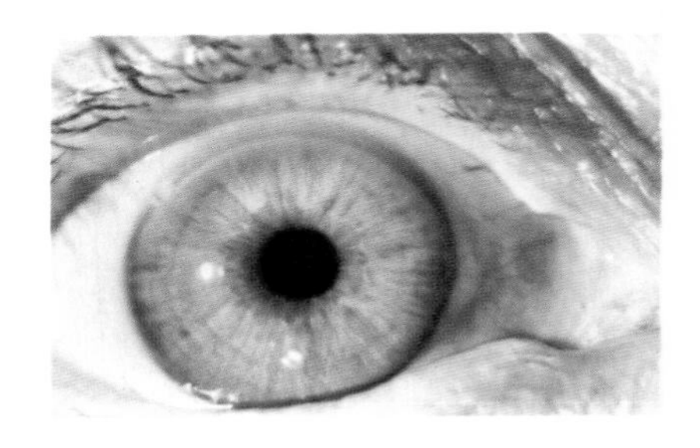

*NO. 21A ANEMIA*
*See Page 43*

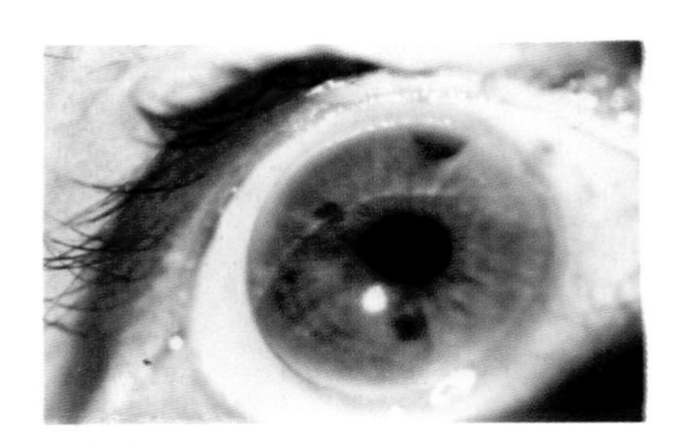

*NO. 21B ANEMIA*
*See Page 43*

*NO. 22 SCURF RIM*
*See Page 43*

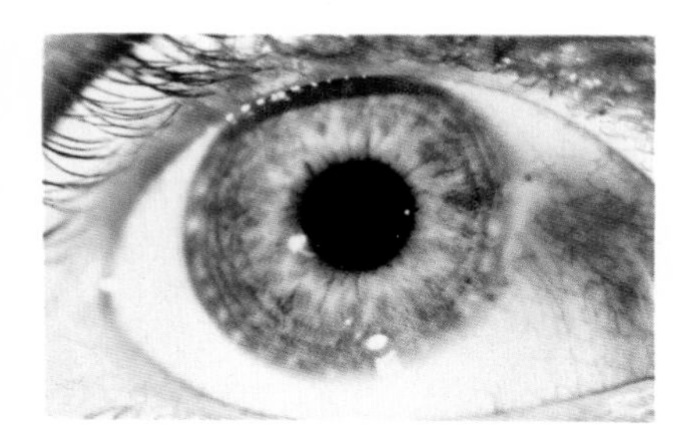

*NO. 23 LYMPHATIC ROSARY*
*See Page 44*

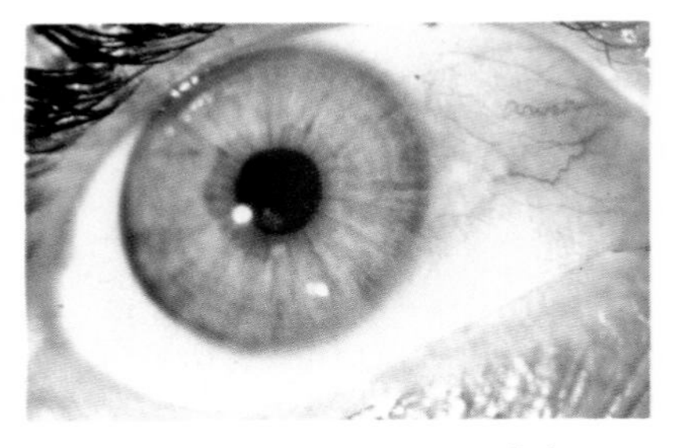

*Adult female – maintains good diet – flat focus; weak vitality; some spots; yellow color; mucus in entire body*

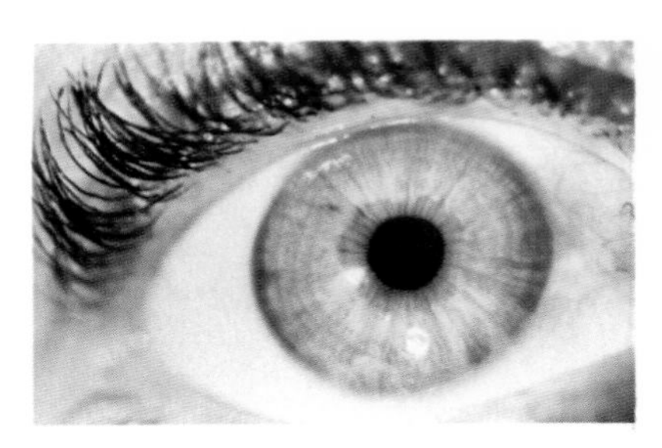

*One month later – after heavy elimination from taking high doses of herbal formula color changes to blue; sharper focus; much improvement*

---

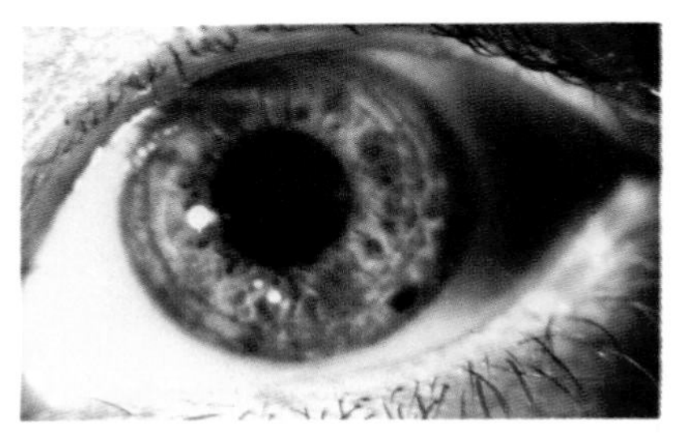

*Adult female – good texture but many spots dark down to lower layers of deterioration*

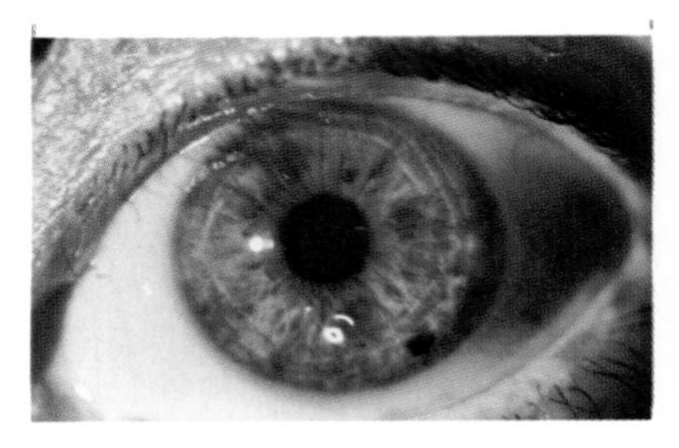

*2 years later using mild food diet and herbs, passed much mucus; spots lighter and smaller; general color improvement; still a way to go*

---

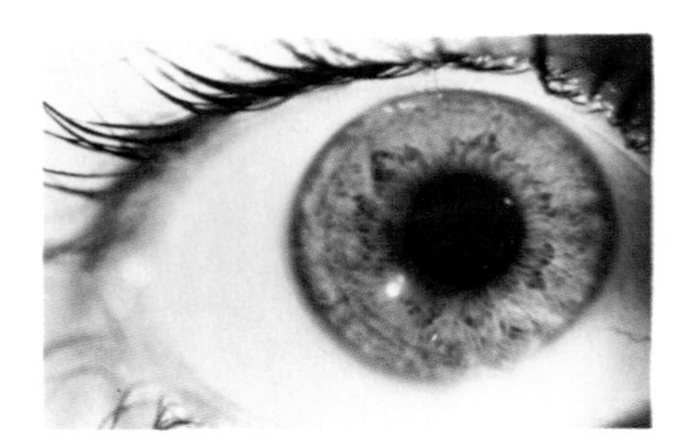

*Adult female – chronic bowel and stomach; lymphatic rosary; weak glands; prolapsed colon; shadow weakness*

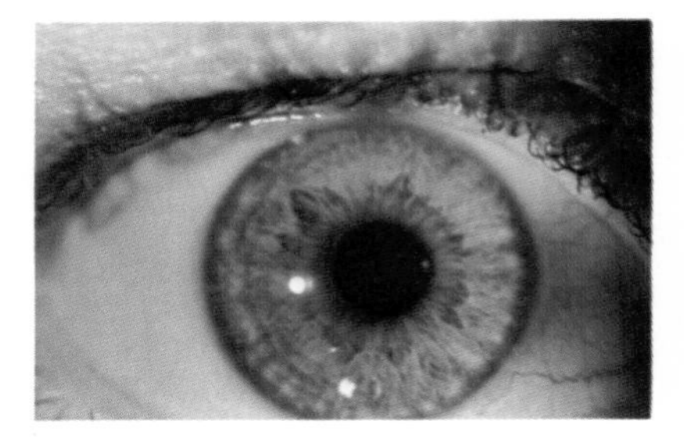

*18 months later, not much diet change but used many herbs – lymphatic rosary whitening; shadows lightening; note area 3:00 almost gone and white*

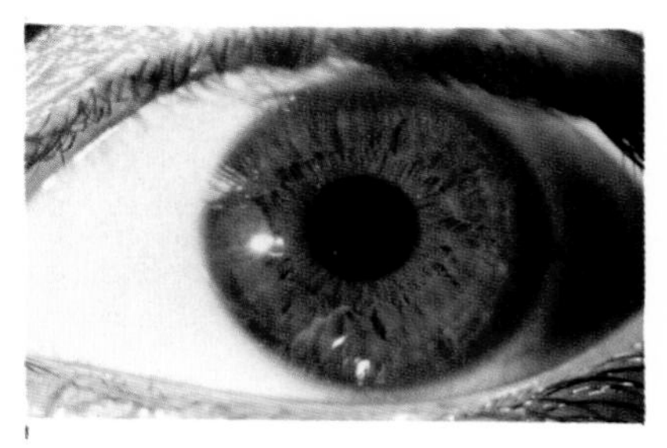

*Adult male – young; parasites; chronic over-all condition serious*

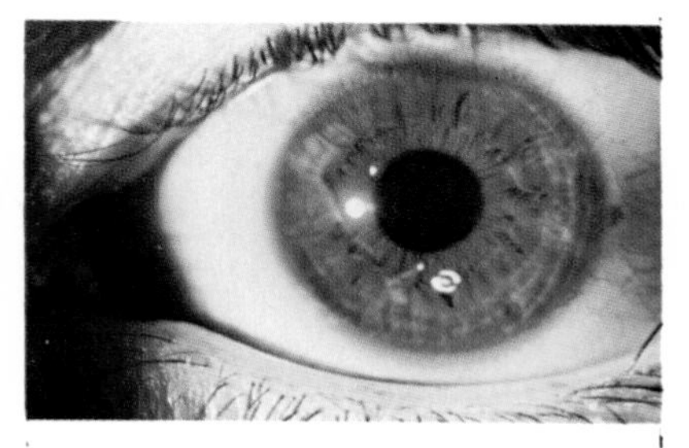

*Stayed on mild food and herbs for 6 months and the color is beginning to change; spots going; parasites going*

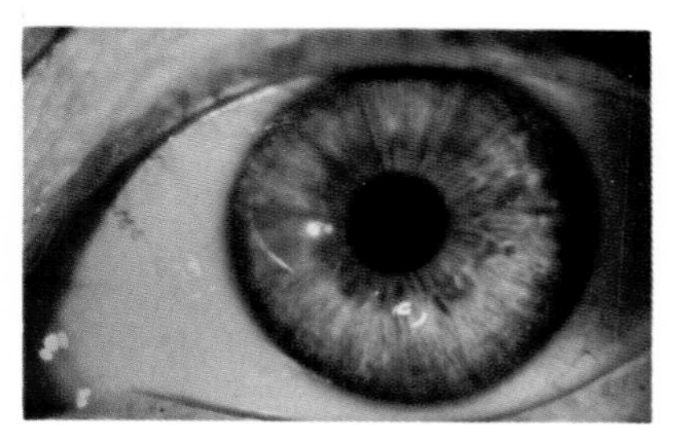

*Adult female – chronic bowel; autonomic wreath; spots; parasites; bowel darkened to lower layer*

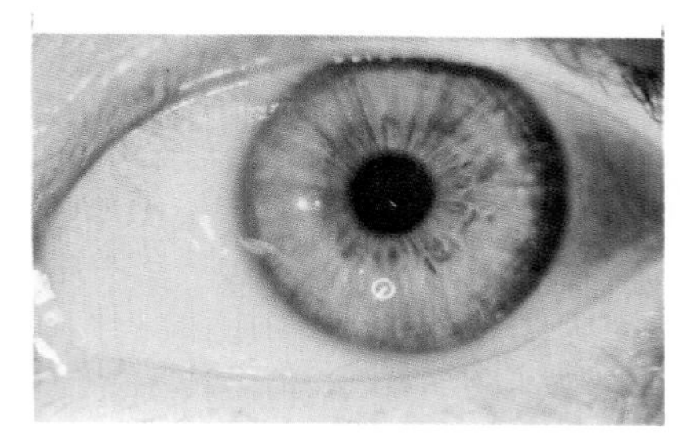

*Note drastic changes after few months mild food diet and herbs*

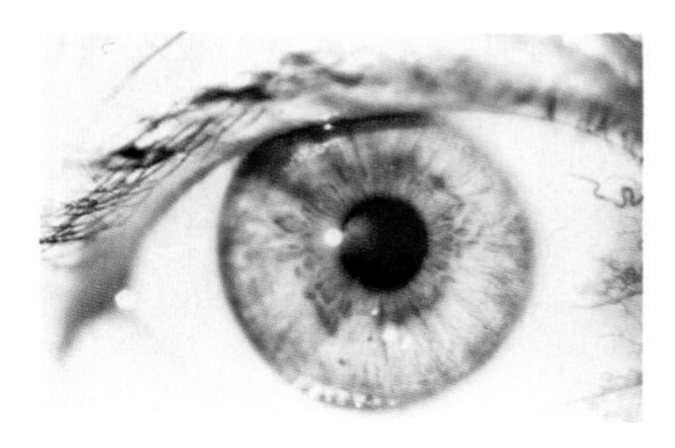

*Was told by her Doctor she had terminal cancer – a year to live*

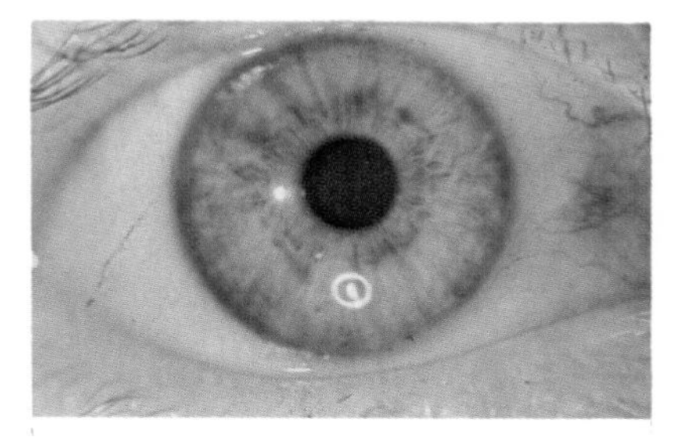

*Note drastic changes after few months mild food diet and herbs*

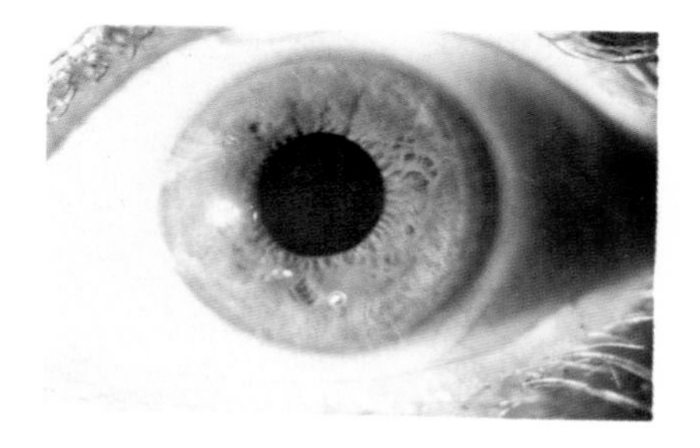

*Serious bowel; chronic bowel and autonomic wreath; chronic mucus in entire body; spots in colon; parasites*

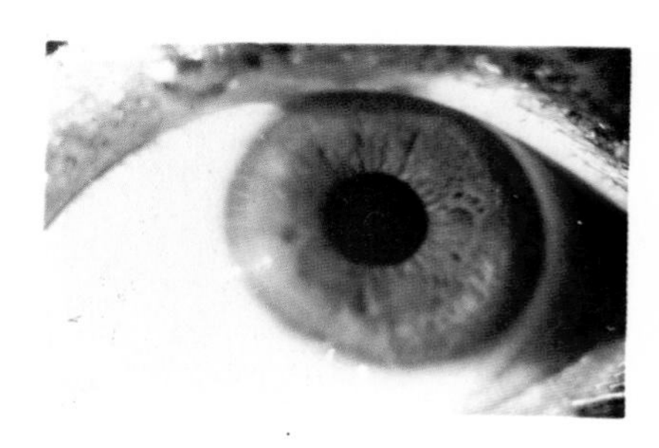

*Some improvement in a few months on mild food diet and herbs – waste begins to gather to colon area as outer areas of the body eliminate and lighten; blue color of eyes begins to show*

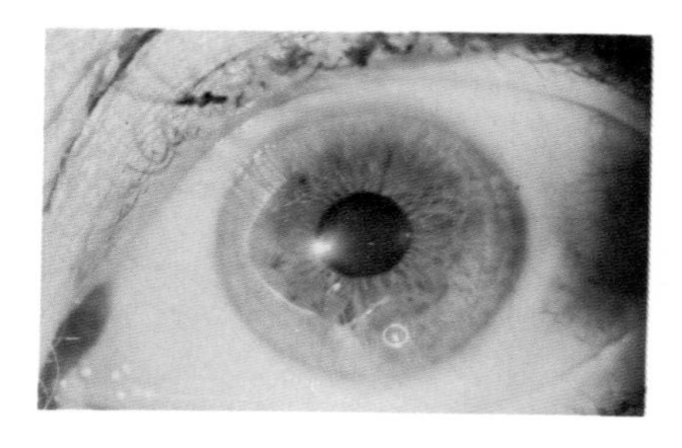

*5 months later very white on improved color area; colon area lightening; heavy elimination; spots going*

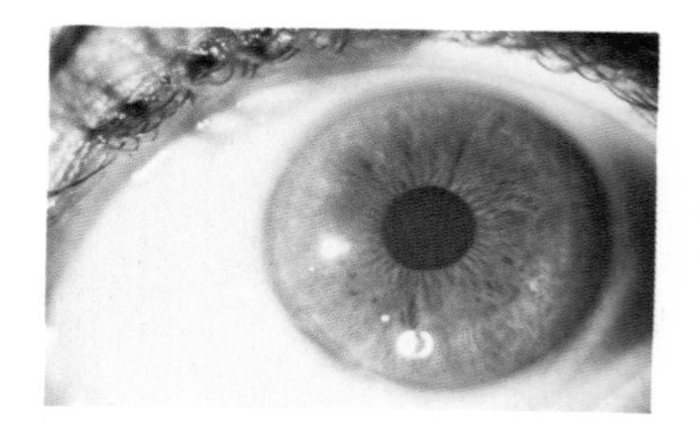

*Color improvement; still chronic in bowel; spots gone; most parasite lines almost gone*

---

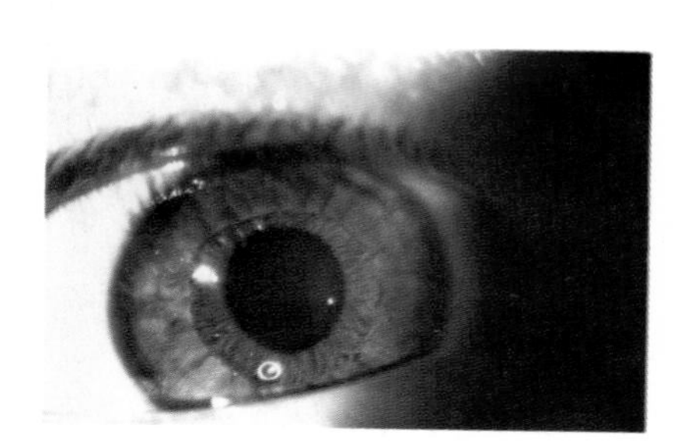

*Adult male – serious over-all chronic condition; parasite; entire weakness flat focus*

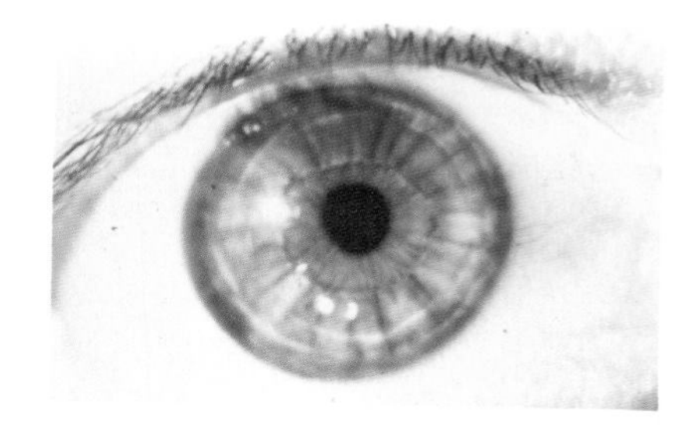

*After several years of mild food diet and herbs – parasite lines show better as other waste is eliminated; sharper focus; increased vitality; under color begins to show; refer to testimony*

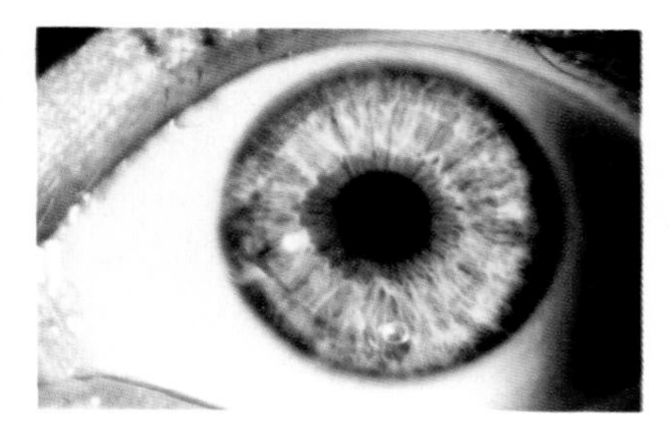

Adult – 2nd picture after several months of mild food diet and herbs; beginning to whiten as elimination proceeds; leisons darkened into lower layers; shadow weakness areas down to lower layers; heavy deterioration; chronic autonomic wreath; shrunken bowel; brown spots; scurf rim

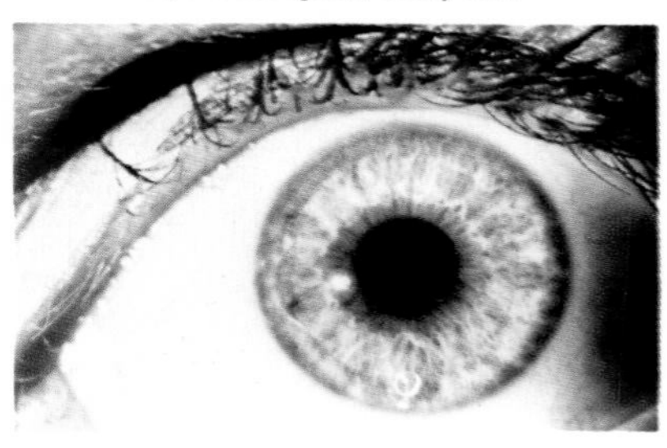

8 months later leisons lightened; eyes whitened; color improved; autonomic wreath improved; scurf rim smaller; nerve ring improved

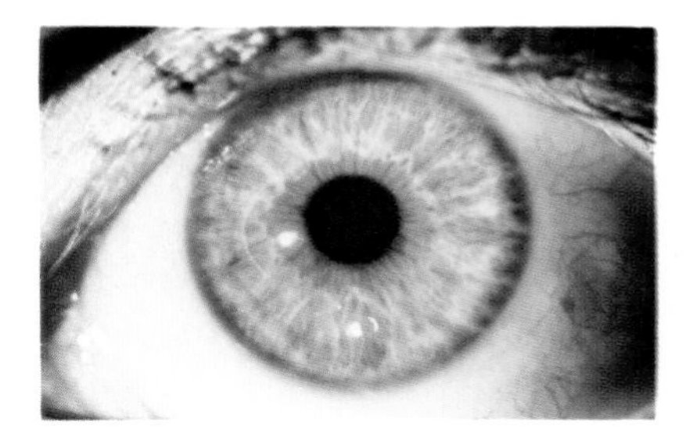

4 months later – all leisons shadow weakness lighter; color brighter; scurf rim improved; heavy nerve ring almost gone; spots gone

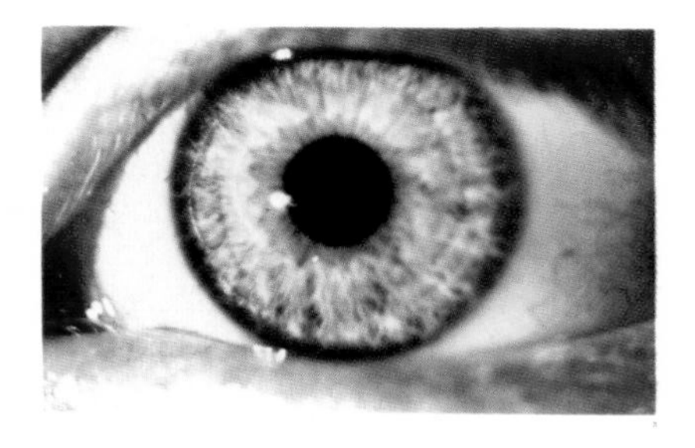

Child – dark leisons into lower layers; infection; lymphatic rosary; bowel spastic with descending colon; acid stomach; some spots; scurf rim; chronic bowel

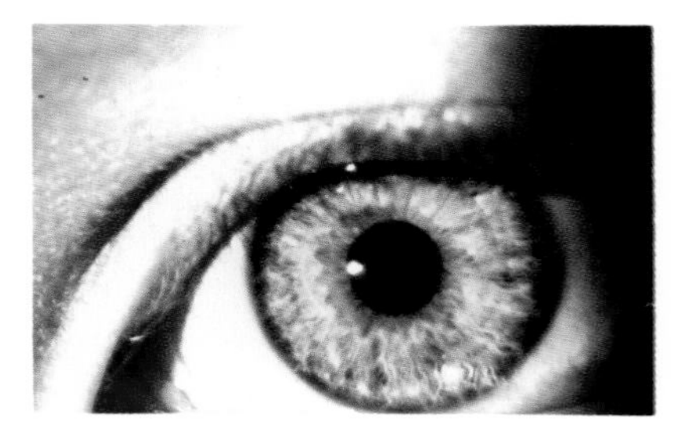

8 months later – mild food diet with some herbs – all leisons lightening; improved scurf rim; chronic bowel improved

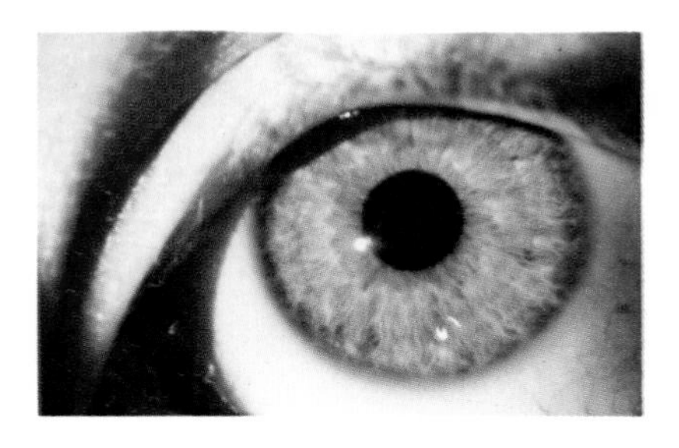

4 months later – brighter color; bowel improved to almost normal; leisons lightened and smaller in size; scurf rim much improved

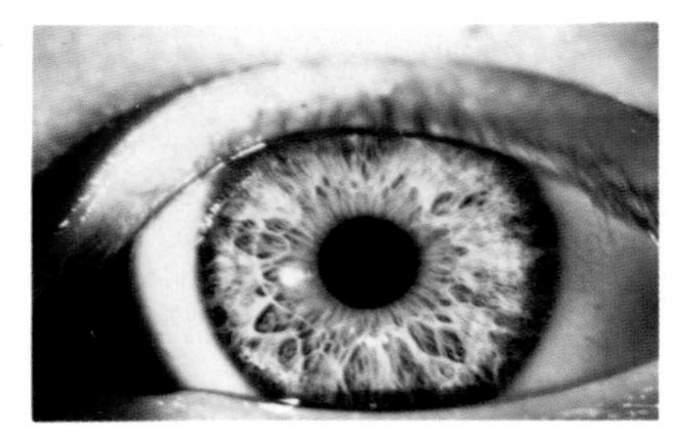

*Child with serious shattered appearance and lesions down to lower layers; spastic colon; scurf rim; heavy nerve ring*

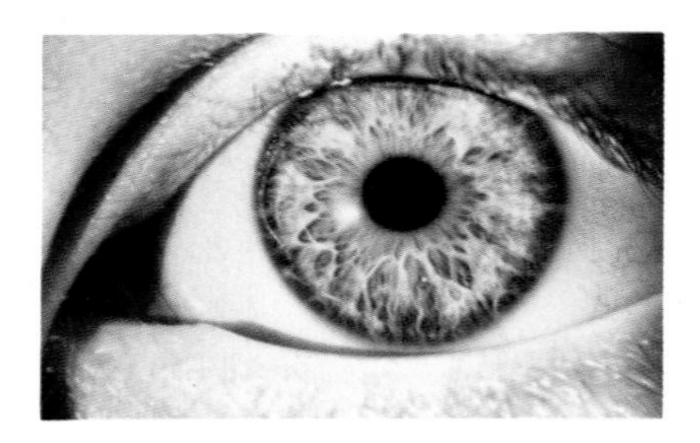

*Mild food diet for 8 months with some herbs; eye whitening; smaller scurf rim; leisons lightening*

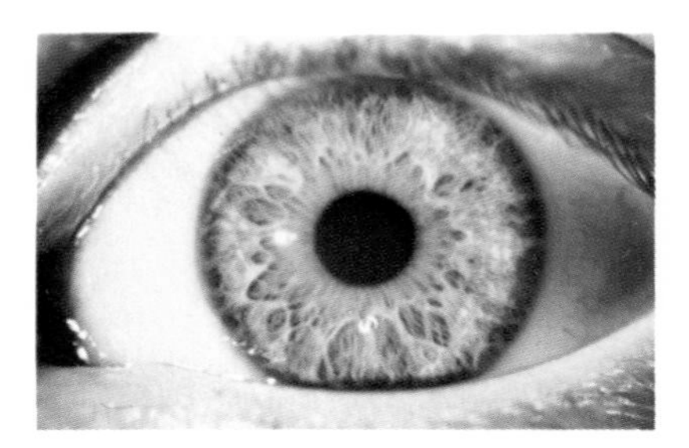

*4 months later – smaller scurf rim; lighter leisons; nerve ring gone; more blue color; eyes still white; elimination continuing with improved colon line*

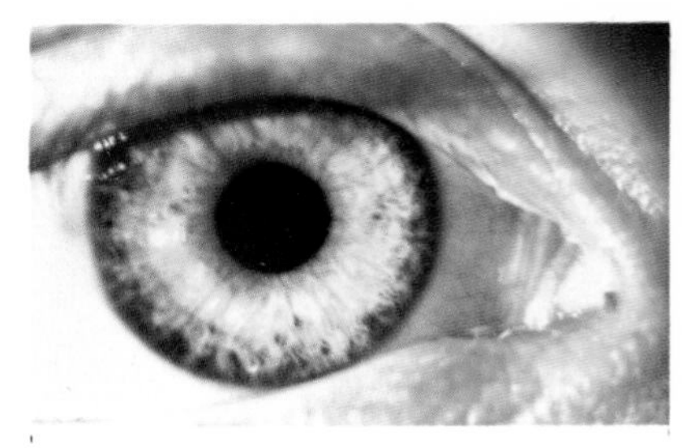

*Child – fair texture; some shadow of inherent weakness; chronic color; acid stomach; parasites in head area; scurf rim wide; chronic and acute mucus around autonomic wreath*

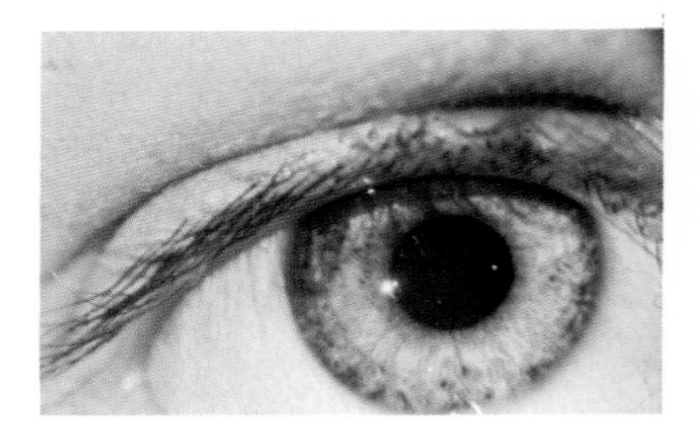

*8 months on mild food diet with some herbs – colon improved; acid stomach gone; eyes whitening; heavy elimination*

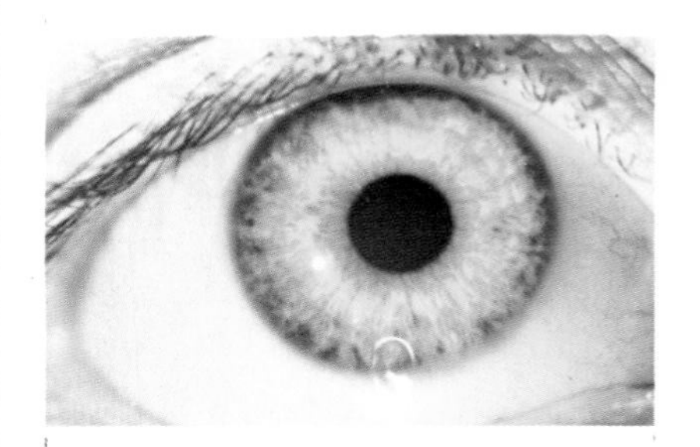

*4 months later – color more blue; still eliminating; parasites gone; scurf rim much smaller*

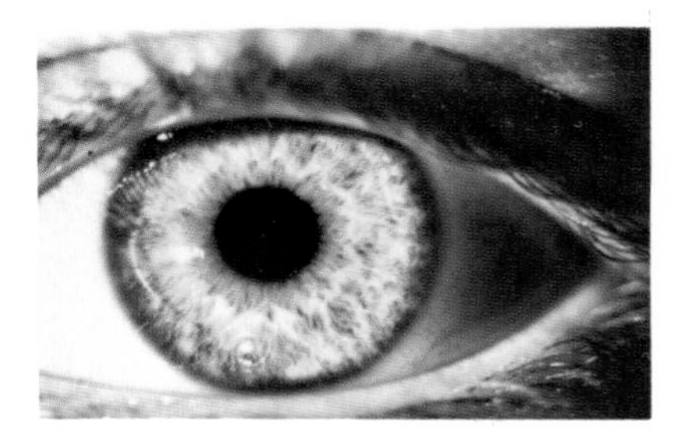

*Child with inherent leisons, darkened down into lower layers – serious*

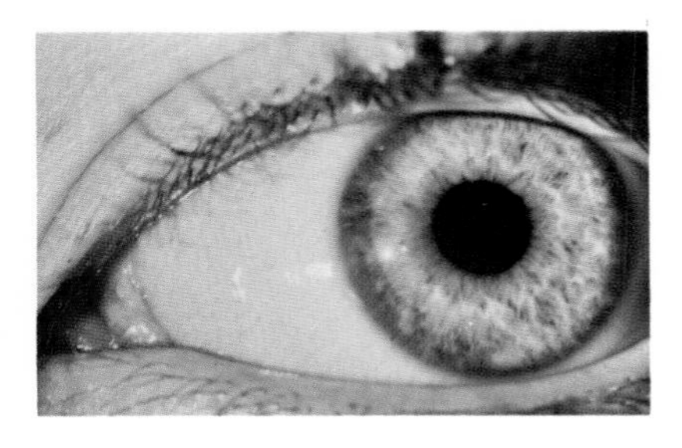

*Mild food diet except school lunch – six months with very little herbs – eyes begin to whiten; leisons lightening*

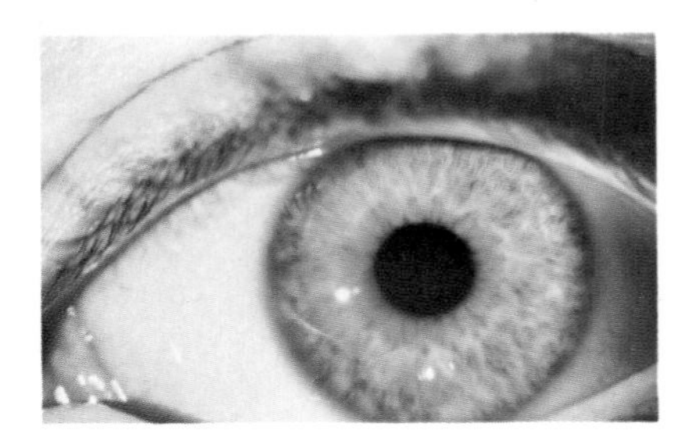

*1 year – continued mild food diet with the exception of school lunch; some herbs – leisons lightening; color blue more pronounced; spots fading*

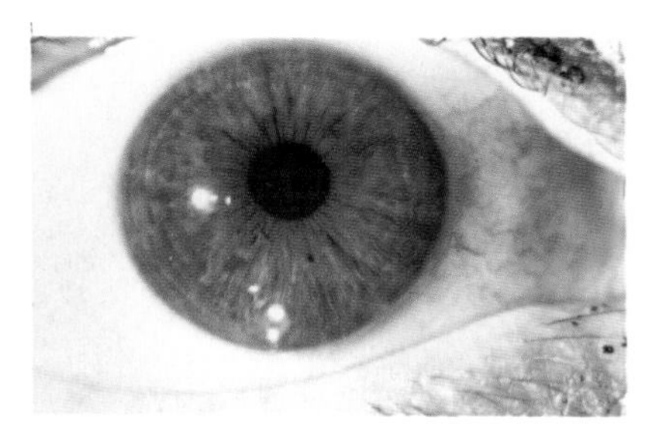

*Refer to testimony – prior to this photo tumor of bowel was passed; tumor also from female area; chronic bowel; autonomic wreath; chronic mucus entire body*

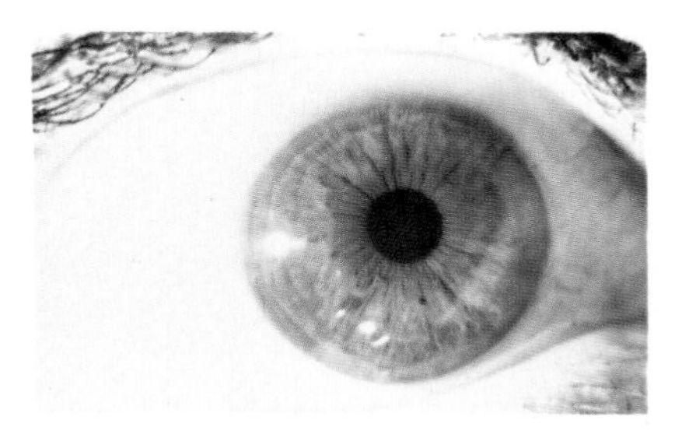

*After healing crisis color changes in eye from brown to blue; still some work to do*

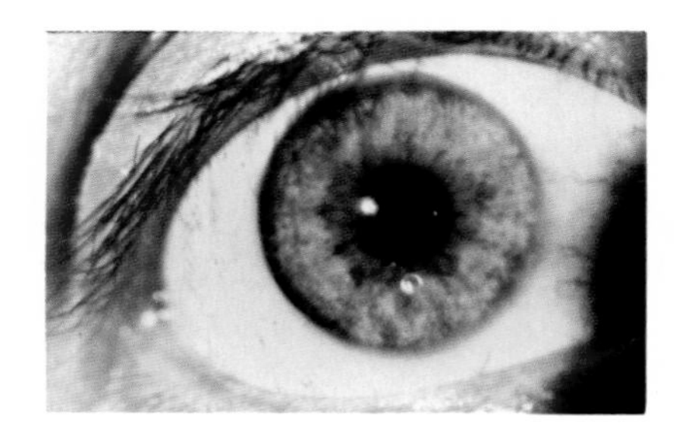

*Colon spots; chronic autonomic wreath; dark inherent weakness shadows; chronic mucus in blood shows yellow cast; parasite lines*

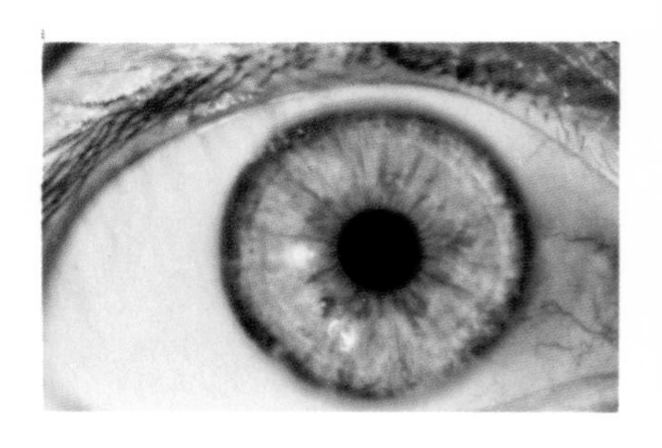

*After few months using herbs, not much diet change, more blue showing and lighter; shadow colon spots diminishing; parasite lines fading to white healing lines*

*The above patient, Jose, has been a patient fifteen years, at the Santa Maria De Tequepexpan Leprosario, Guadalajara, Mexico. This was the day we started treatments, amid the doubts of the doctor and padre, but we had God on our side, and it worked!*

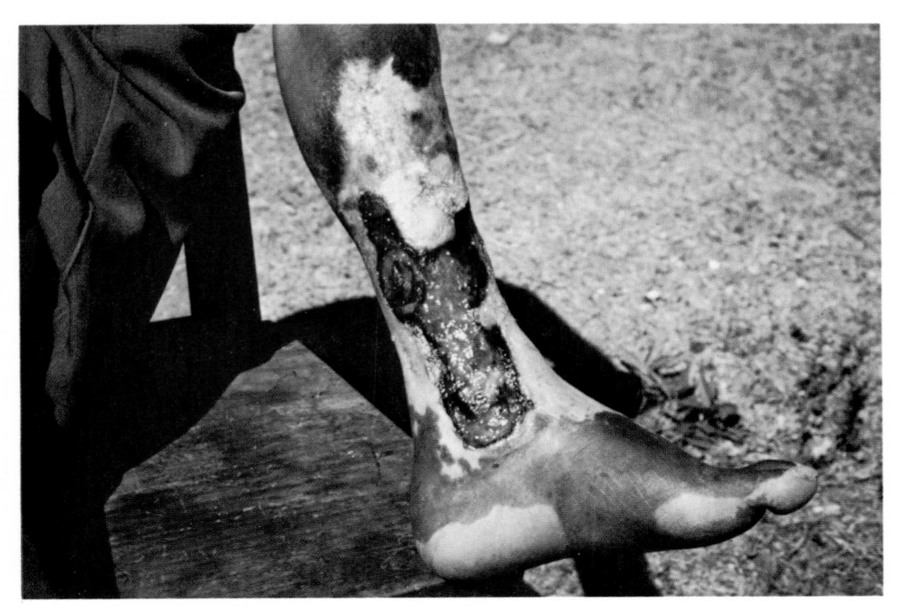

*Eight weeks later, a vast amount of improvement. The foot is no longer withered, and the calf, after fifteen years of disintegration seems to be forming near normal. The new, pink healthy flesh, is converging from both sides and the top, on the disappearing lesion. In another eight to ten weeks, the patient will be dismissed and permitted to go home.*

See Page 17

# IRIDOLOGY CHART

**1. Texture:**

| Excellent | Good | Fair | Poor |
|---|---|---|---|
| Fine | Some lesions, not chronic. | Lesion, darkening. | Shattered, darkened lesions |

**For Desired Change:** If inherently poor texture usually not possible to make a change but must live a more healthy life than those having a finer texture, in order to survive and remain well.

**2. Focus lines:**

| (E) | (G) | (F) | (P) |
|---|---|---|---|
| Sharp. | Medium sharp. | Flat. | Out of focus. |

**For Desired Change:** Can be changed in many cases by eating more live foods, especially in old age. When the old go to a softer diet and fewer raw foods, dark chronic color is increased. When many raw juices are added to their diet, the strength will increase and focus will sharpen where focus is already flat, showing weakness. More raw food and vitamins will sharpen the focus increasing endurance and strength.

**3. Heart:**

| (E) | (G) | (F) | (P) |
|---|---|---|---|
| No lesions or markings of any kind. | No wreath involvement. No lesions, excursion of autonomic wreath into heart area. Would still be all right. Must be corrected. | Inherent lesions, no spots, no involvement of autonomic wreath. | Lesions and spots or involvement of autonomic wreath. |

**For Desired Change:** Usually cleaning colon and entire body relieves the burden on the heart, but too much toxic waste moving too fast pumping through the heart as the body expels, can cause more damage, like taking a shot of deadly poison. People with heart problems must move slowly, eliminating with mild food and cleansing herbs and must not try too fast or go on fruit only. Herbs that cancel out poison, such as saffron and dandelion, should be used. Then heart herbs such as Hawthorn (to strengthen and rejuvenate the heart), Potassium gluconate (to strengthen the heart muscle) and vitamin E may be taken. One of the most deadly poisons to hurt the heart in the process of elimination is uric acid. Where kidneys are not properly carrying off the waste, dandelion and saffron have a way of neutralizing this uric acid. Other kidney herbs help to activate the kidneys to a better function. Potassium gluconate and celery juice also help to pull off fluid. Where the kidneys will not respond to such methods, the bowel must be forced to evacuate to a fluid state by highly laxative herbs. This takes the burden temporarily off of the kidneys and causes waste to flush out rather than circulate through the heart.

**4. Chronic Color-toxicity, spots or shaded:**

| (E) | (G) | (F) | (P) |
|---|---|---|---|
| Natural color of eye. | Beginning to color into yellow or light brown. | Brown spots, light brown in lesions. | Lesions deteriorating to lower layer black. |

**For Desired Change:** Any chronic spots or lesions show deterioration, mild food and cleansing herbs until spots are removed and healing is complete. Often people feel better; so stop and go back to concentrated foods before they have sufficiently cleansed.

**5. Colon and Stomach:**

| (E) | (G) | (F) | (P) |
|---|---|---|---|
| Normal colon stomach lines outlined only by darker shades than eye color. | Some deterioration, no chronic color. | Distention of incursion to pupil, darkening spots or lesions. | Distention or incursions to pupil, dark spots or lesions, parasites or |

**For Desired Change:** Enemas, cleansing herbs, mild food diet, parasite herbs, psyllium, slippery elm (for ulceration or soreness, fissures or hemorrhoids). Cayenne and golden seal for any bleeding; vitamin B6 or acidophilus.

**6. Autonomic Wreath:**

| (E) | (G) | (F) | (P) |
|---|---|---|---|
| Outlining, only a darker shade than color of the eye. Medium thick- | Good thickness, white or light brown in color. | Uneven thickness, darkening in color. | Uneven thickness or very thin, darkening color, splitting separations. |

ness all the
way around.

**For Desired Change:** Nervine herbs (see *Is Any Sick Among You*), vitamin B. complex, calcium lactate or gluconate, cleansing herbs, mild food.

**7. Nerve Rings:**

| (E) | (G) | (F) | (P) |
|---|---|---|---|
| No rings showing. | Small and few showing. | One or two disconnected rings, white in color. | Brown rings two or more, three or more white rings. |

**For Desired Change:** Where chronic showing, check nerve wreath also. When white nerve rings come and go, find out who or what is causing the nervous tension. Change environment or attitude towards environment. B complex, calcium lactate or gluconate and nervine herbs.

**8. Glands:**

| (E) | (G) | (F) | (P) |
|---|---|---|---|
| No marks of lesions on gland areas, straight lines of iris. | Possible wavy lines but no markings or lesions on gland area. | Lesions on glands, no deterioration, wavy lines. | Lesions, wavy lines, deterioration on lesions or spots appearing on gland areas. |

**For Desired Change:** Open lesions are easier to change because waste has a way out. Mild foods and cleansing herbs.

**9. Open Lesions; extent of deteriortion, determine which organ:**

| (E) | (G) | (F) | (P) |
|---|---|---|---|
| None showing. | Small lesions only a shade darker than iris. | Few larger lesions with some deterioration showing. | Many lesions, much deterioration, brown and black. |

**For Desired Change:** Open lesions are easier to change because waste has a way out. Mild foods and cleansing herbs.

**10. Closed Lesions: extent of deterioration determines which organ:**

| (E) | (G) | (F) | (P) |
|---|---|---|---|

Same as above only not so serious as a closed lesion.

**For Desired Change:** Same as above but more difficult to change. Healing lines will appear across lesions as if you were weaving criss-cross threads over the lesions.

**11. Parasites:**

| (E) | (G) | (F) | (P) |
|---|---|---|---|
| No chronic deterioration, no radii solaris lines or radiation lines. | Any chronic disease, radii solaris lines. | Any chronic disease, radii solaris lines, radiation lines. | Heavy showing of chronic disease, radii solaris lines, radiation lines especially on animation area. |

**For Desired Change:** Where cancer parasites (progenitor cryptocide) cancer formulas, sometimes additional herbs such as chapparel and black walnut. Where Salmonella and trichinosis, pumpkin seed (also see herbal pumpkin formula in *Is Any Sick Among You*)Where Lupus, malaria or leprosy: black walnut. Lines will fuzz off then go to white, then disappear as parasites are killed.

**12. Injury:**

| (E) | (G) | (F) | (P) |
|---|---|---|---|
| No injury. | Injury from infection adhesions. | Some, of a a minor nature. | Brain damage from stroke or other heavy injury. |

**For Desired Change:** For injury causing scar tissue, vitamin E. Must live on a better than average diet, depending on damage done.

**13. Surgery:**

| (E) | (G) | (F) | (P) |
|---|---|---|---|
| No surgery. | Minor surgery. | Scar tissue, adhesions, minor surgery brown or black. | Major surgery black and brown spot remaining even after cleansing diet. |

**For Desired Change:** Scar tissue, same as above. When permanent damage has been done, area will remain brown or black.

**14. Drugs:**

| (E) | (G) | (F) | (P) |
|---|---|---|---|
| No drugs showing. | Some small sodium rings in one area. | Sodium ring medium. | Large Sodium ring and drug red ring around pupil. |

**For Desired Change:** Drugs can be expelled as the body is placed in a fasting or semi-fasting state. As the move into circulation, palpitations will often occur. Care must be taken in the elimination process when many drugs have been taken of an extremely toxic or narcotic nature. Taking care of the heart as has been mentioned, mild foods and cleansing herbs. (see heart).

**15. Acute:**

| (E) | (G) | (F) | (P) |
|---|---|---|---|
| No white showing. | Small amount of white showing. | When eliminating, large amounts of white showing. | When not eliminating, and large amounts of white showing. |

**For Desired Change:** Acute inflammation or acute illness when eyes show white over the top of the iris color, waste is on the move in the blood, help nature get the waste out of the body. Fruit only, cleansing herbs, laxative herbs and high doses of vitamin C and enemas.

**16. Liver, Gall bladder:**

| (E) | (G) | (F) | (P) |
|---|---|---|---|
| No markings showing. | Shadow darker than iris color indicated weakness. | Liver or gall-bladder, either showing chronic color. | Both liver and gall-bladder shows chronic color. |

**For Desired Change:** Mild food and cleansing herb formula. Gall bladder purge (see *Is Any Sick Among You*). Golden seal, saffron, lemons and raw beet juice all activate and clean the liver.

**17. Kidneys:**

| (E) | (G) | (F) | (P) |
|---|---|---|---|
| No marking on kidney | Shadow deeper color than eye. | No deterioration or white. | White, infection or brown or black showing deterioration. |

**For Desired Change:** Where infection, same as with acute. Where deterioration cleansing herb formula, mild food, uric acid removing herbs such as saffron, dandelion. For healing, juniper herb or juniper oil, and cornsilk.

**18. Bones — Hypercalcemia osteoporosis:**

| (E) | (G) | (F) | (P) |
|---|---|---|---|
| No markings showing. | No markings showing. | Inflammation of cornea possible beginning. | Pit-like appearance on body structure (osteoporosis), gray, glandular appearance either side of pupil. |

**For Desired Change:** Gland involvement, check all glands, related, mild food and cleansing herbs.

**19. Eyes:**

| (E) | (G) | (F) | (P) |
|---|---|---|---|
| No markings. | No markings. | Shadow inherent showing or white showing infection. | Deterioration showing chronic in color. |

White showing on the eye area would be inflammation. In order to move inflammation all fruit, herb laxatives, enemas and high vitamin C. Where deterioration is showing, mild food diet, cleansing herb formula, vitamin B1, B2 and vitamin A. Parsley is one of the richest natural sources of vitamin A, even higher than carrots.

**20. Ears:**

| (E) | (G) | (F) | (P) |
|---|---|---|---|
| No Markings, | No Markings. | Shadow inherent showing or | Deterioration, showing chronic |

| | | white, showing infection. | in color. |
|---|---|---|---|

**For Desired Change:** White would be infection or inflammation, where it is earache an ice bag to scatter infection away from ear area. Enemas, fruit diet and laxative herbs which will drag the mucus out of the body. If it is a fungus, mullein herb in a tincture or strong tea or pumpkin seed oil, pumpkin seed or black walnut internally. Often where there is chronic deterioration showing, a parasite line will show also, or a radiation line. Before the mucus so involved can be moved out the parasites must be killed. Black walnut and pumpkin seed.

**21. Anemia:**

| (E) | (G) | (F) | (P) |
|---|---|---|---|
| No anemia showing. | No anemia showing. | Infection, or acute showing too often when not on a cleanse. | Anemia showing anywhere in eye. |

**For Desired Change:** Eye is white most of the time, where not on a cleanse, this indicates too much mucous in the blood, can lead to leukemia. Stop intake of mucous-forming foods and treat as a cold. Anemia in small areas of the eye, which looks like the peripheral lines, become obscure; these places indicate too much mucous in the area and a lack of oxygen. If it shows in the brain area, more calcium is necessary to bring oxygen to the brain as well, as a good cleanse.

**22. Scurf Rim:**

| (E) | (G) | (F) | (P) |
|---|---|---|---|
| No scurf showing | No scurf. | Small, irregular showing scurf, no deterioration. | Wide scurf, deterioration |

**For Desired Change:** When the scurf rim shows only a darker color than the color of the eye, this is weakness and will result in an added burden on internal skin as well as bowel and kidney stimulation of the skin. Steam or sweat baths, sun baths, massage or bath brush and a good towel rub. When chronic, mild food, cleansing herbs along with those things listed above.

**23. Lymphatic Rosary:**

| (E) | (G) | (F) | (P) |
|---|---|---|---|
| No markings. | No markings. | Few white | Rosary extend- |

| | |
|---|---|
| markings in various places on rosary. | ing all or much of the circumference of the 6th zone. Chronic in color, even in beginning stages. |

**For Desired Change:** Can develop into Hodgkins disease even if the rosary remains white for too long. Clean entire lymphatic system. Mild food and cleansing herbs, particularly laxatives. Flush the body lymph. When chronic cholesterol, lecithin helps to emulsify and loosen fatty wastes in arteries. Check for lymphatic rosary across bony structures indicating arthritis. Check glands areas, incorrect calcium balance can cause further calcium buildup, hardening arteries.

## SUGGESTED WORK CHART

**Excellent Good Fair Poor Suggested remedy — Remarks**

1. Texture:
2. Focus:
3. Heart:
4. Chronic Color — toxicity:
5. Colon and Stomach:
6. Autonomic Wreath:
7. Nerve Rings:
8. Glands:
9. Open Lesions:
10. Closed Lesions:
11. Parasites:
12. Injury:
13. Surgery:
14. Drugs:
15. Acute:
16. Liver — Gallbladder:
17. Kidneys:
18. Bones:
19. Eyes:
20. Ears:

21. Anemia:
22. Scurf Rim:
23. Lymphatic
    Rosary:

**Mild Food to Cleanse:**

All vegetables and fruits (as much raw as possible)
Fruit juice (canned, raw or frozen)
Soft oil (raw, cold pressed)
All nuts (must be raw)
Honey (raw)
Sprouts (alfalfa, bean, grain)
Seeds
All starch vegetables must be baked: squash, potato, parsnips, yam

**Some Herbs that Detoxify:**

Echinacea
Gentian
Catnip
Bayberry
Golden seal
Iris moss
Fenugreek
Chickweed
Comfrey
Prickley ash berries
St. John's wort
Blue vervain
Red clover
Saffron
Dandelion
Mandrake
Senna
Cascara sagrada

**Some Herbs to Kill Parasites**

Black walnut
Pinkroot
Pumpkin seeds
Pulsatillo
Borage
Mandrake
Tansy
Cyani flowers
Yellowdock
Bearfoot root
Poke root
Nettle
Heal all

**Semi-fast:**

To cleanse in time of famine or limited food supply and still gain adequate nourishment.

1 bowl of mush a day or
1 glass of milk or
1 slice of bread (whole wheat).

# CHAPTER THREE

# WHAT'S IN A TESTIMONY

*Open minds and listening ears are ever eagerly searching for truth and ways to improve and truly enjoy life.*
*Closed minds and stopped ears are forever locking their doors in the faces of opportunity, progress and peace. Thus do they doom themselves to stagnation, decay and misery.*

*J. Melvin Gibby*

*8/13/61*

Like health, happiness, and a good marriage, a testimony is a fleeting thing: you have to work at it to keep it. For brief episodes as we apply what we have learned, our testimonies of any law — from putting our finger in the socket and experiencing a shock, to the profound anguish resulting from hateful words — exercises a certain influence on us each living moment. It is, therefore, important that we maintain a good memory bank in our computer brain in order that we may avoid too many mistakes. Even though most mistakes only hurt a little, others can kill us. Too many of us live naively, inattentive to those things that bring us happiness or pain either in our physical body or in our surroundings. If we were lost in a jungle with wild animals and physical danger, we would become as cunning as a cat, we would give everything we did a little more thought. We would become more aware, logical, and emotional — feeling, thinking, watching, praying, and then acting more correctly for each given situation. Too often when a person has had a beautiful testimony of natural healing methods, the passing of time and a busy schedule lets them slip back into bad habits, and they even forget what they did to get better so as to avoid further problems.

> *Every hardship we encounter in life, if conquered, is a great stepping stone toward our ultimate goal whatever it may be. But unconquered, it remains a persistent stumbling block, which stubbornly impedes our progress until we develop the will and the intelligence to remove it from our path.*
>
> *J. Melvin Gibby*
> 9/6/61

A so-called learned man speaking on a talk show recently told how you could tell a quack medical book. He said it would have testimonies. This statement reminds me of Paul before King Agrippa when Festus said, "Much learning doth make thee mad." Though Paul was not mad, sometimes much learning seems to make some men of science mad. They forget that testimonies written in the many books of the past have prevented us from re-living the dark ages. All books are testimonies of someone's experience — good or bad; these are the things we learn from. It is our task to search these things out, weigh carefully each witness, and give our whispered prayer to heaven for confirmation, and then proceed accordingly. As you read these testimonies, it will be your prerogative to make a judgment.

> *Truth is any real and eternal principle that is inalterable by man, time conditions or even by God Himself.*
> *Error is a departure in thought, word or deed from the principles of fundamental truth.*
>
> *J. Melvin Gibby*

It should be noted that these people are from different religious backgrounds and yet the thread of faith in God seems to weave its way through most of their testimonies in their search for truthful answers to their health problems. Far be it for me to take the credit some of these people give to me in these testimonies. All I have done is teach the truth, they picked it up, applied it and proved it, at least to themselves. For my part I thank the Lord for the knowledge He has given me and the glorious testimony He has given me.

My gratitude also reaches out to all these wonderful people who have had the courage and faith to find these answers.

**Ruth Jacobson**

Licorice root has been an invaluable aid to me in the treatment of low blood sugar. It has enabled me to carry on throughout the day without the low blood sugar letdown, even if I miss a meal. It has eliminated the fuzzy mind when I just couldn't think, concentrate, or remember. I sleep better, no longer have the "go to bed tired, get up tired" feeling. Now I carry licorice root with me instead of nuts. If I happen to have forgotten to take it, or if I encounter severe stress, a dose of licorice gives me a quick pickup and new energy within a few minutes. I find that diet plays an important part in the effectiveness of licorice root. White sugar and white flour products can wipe out the benefits of the herb. I took time to test this out thoroughly before making my statement. I wanted to be sure I was accurate on it. I just started my son Darrell on it as he was complaining of exhaustion and he is getting the same pickup I do with it.

**Myrth Kidd**

I would like to add my testimony to those who like mine have read and put to the test, Mrs. Griffin's book, *Is Any Sick Among You*". I am a mother of ten children, all of them of average good health, at least I have felt this up to our last child. Even though to most people, she seemed to follow a healthy pattern, because she ate well, and was big for her age in stature and weight. This is a mistake many of us make, but my mother's instinct told me all was not well, especially when throat and ear infection became a common thing to her. I knew also, her little body was sluggish, and that her chubbiness was not a good sign of healthiness. I pondered these things in my mind. The doctor told me, and I'm sure he was truthful as far as his knowledge and instruments could tell, that she was fine. So on with the food and anti-biotics with some kind of "give her a few years and she'll outgrow it" thought in my mind we went steadily on. About six months before I heard Mrs. Griffin's first lecture, I picked up another book at the health food store, and after reading it, I decided it would be good for me to try some juice fasting and exercise, to sum it all up, try and pull myself together, after all I was forty-six with ten children and, I felt every bit of it. I was amazed how soon I began to see results. It was spring, and as we gardened and gathered asparagus, the truth my mother taught me, for she was a nurse in days when only the herb and natural foods were their medication, "that God not man produces that which is good for the stomach". That spring and summer prepared me for one of the greatest experiences of my life, the lectures and "the book", as we refer to it at our house, by LaDean Griffin. Not only has it brought a new way of life to me, but my little Heather, whom I spoke of earlier, is like a new child. What a joy it has been to watch her change. No more bottles of anti-biotics, because in studying her eyes, I could see her weaknesses. Oh, at first this was frightening, for I could see my fears did have merit, all was not well with her, but then we got with the program, no animal products, no white sugar, no white flour, in fact straight fruit, vegetables, nuts, raw honey and herbs. I watched the change begin to take place both in her eyes and her whole self, she was almost three years old when we started and now some sixteen months later I can hardly believe the difference. She still has her inherent weaknesses, but we know this now, and we also know that a healthy, happy life can be hers, if we do not add to her problems. I could tell you many more things about myself but space won't allow, so to those who read my testimony, I bring to you good tidings, it's a great life and because of an old but new knowledge, I am convinced we can enjoy it to its fullest.

**LaVerne A. Milam**

Herbs working from within the body; chiropractic from without: Blood pressure record of LaVerne A. Milam, age 57, beginning 4/21/75. Record on file in office of a doctor in Alhambra, Ca.

| | |
|---|---|
| 4-21-1975 | 160/100 |
| 5- 1-1975 | 150/90 |
| 5- 8-1975 | 140/90 |
| 5-13-1975 | 135/80 |
| 5-15-1975 | 130/80 |
| 5-22-1975 | 130/80 |
| 5-28-1975 | 130/80 |
| 6- 3-1975 | 130/80 |
| 6- 6-1975 | 130/80 |

| | | |
|---|---|---|
| 6- 9-1975 | 120/80 | |
| 6-13-1975 | 120/80 | |
| 6-16-1975 | 120/80 | |
| 6-20-1975 | 130/80 | (after treatment, down to 120/80) |
| 6-26-1975 | 140/80 | (after treatment, down to 130/80) |
| 7-10-1975 | 120/80 | |

Beginning on 4-22-1975, I knew the combination of herbs for high blood pressure was capsicum and garlic, but by mistake I picked the heart combination of hawthorne berries and capsicum off the shelf in the dim light and without my glasses, and started on a plan of two, four times a day; at meal time and just before bedtime. On my second trip to the doctor, as you can see as of 5-1-1975, it was down. But the doctor was not satisfied. He asked whether I was on the herbs which I had told him I was going to try. I told him, " yes, but it's like this, although I should be having two, four times a day, today for instance, I had two at breakfast, but I wasn't home for lunch, I won't be home for dinner, and, in fact, I won't even get home until sometime around twelve tonight." He then informed me I had to do something definite about this, because he wanted it to come down. So I told him, alright I would commence to carry a day's supply in my purse and faithfully use it. Before the next visit to his office, I suddenly discovered I had been taking the wrong combination, but you will notice, since it was for the heart and I had half the correct combination by having capsicum, the pressure had started down. At this point I would like to explain that all my life I have never had a problem with high blood pressure and couldn't believe it when the doctor told me this was high. I really felt tops. I had never had any of the symptoms or warnings most people have; such as, flushed face, tingling in hands or legs, tinglings in the head or neck. I didn't even know what the normal for me should be. I had never had to be concerned. So, the next visit on 5-8-75, you can see it is still dropping. This was when I asked the doctor what my normal blood pressure should be. So he showed me how you start with the average twenty-year old pressure of 120/80, and for each two years of age after twenty, you add one MM to the reading, which made mine for my age to be 138mm/90mm. As you can see, my reading for that day was 140/90, about a week and a half after the first reading. Then it occurred to me and I asked the doctor, "Doctor, if all systems in your body are go and you are really feeling great, why should age even enter into the picture? Why shouldn't I be able to reach the level of a twenty-yr. old?" He just grinned, and said, "Listen, you just stay at 138/90 and I'll be happy." As you can see by the above chart, approximately six weeks after the first reading, I reached the reading of a twenty-year old and the doctor was amazed.

**Mabel U. Ziros**

I wanted to add my testimony for herbs. The first one I tried was black cohosh one year ago. I have had a complete historectomy and one fourth of one breast removed because of fibroid tumors, and have had to take hormone shots. I stayed off hormone shots as long as I could and when I went back for a checkup I was low in hormones and the doctor put me back on them and then after awhile he started me on the pill and after awhile I started getting the lumps in my breasts, very similar to the problems I had before surgery and I felt very discouraged. I was telling my problem to a friend of mine who was selling herbs and she had your new book, *Is Any Sick Among You*". She said, "Let's check in this book and see which of the herbs is the female hormone. We looked it up and it is black cohosh, and after reading about it I decided to try it, there was nothing to lose, the cost was very little compared to the shots or the pill. I did not know how many capsules to take but this being natural, I was not afraid to take two capsules morning and night, and sometimes at noon. The results have been unbelievable and I do not have the lumps in my breasts, or any of the side effects I had with the shots and the pills. I did not even finish the bottle of pills after I started using the herbs. Mrs. Griffin, before I had a terrible problem with dryness and the doctor said all I could do was use KY jelly, well now I do not have to use that and I am much more comfortable. So I am thankful to my friend and to you for the knowledge I have gained from your books and your lectures on herbs. Now I am using the various herbs for various problems and I have had equally good results. The more I study and learn about herbs the more thankful I am for the convenience of the herbs we can now get in the capsules.

**Elizabeth B. Silver**

Having been plagued with psoriasis since high school age and having tried various treatments through the years, I finally, (at age 57) found notable relief in the herb and juice program of Warren Kirley Crosby. I am convinced that following this regime periodically will counteract the flare-up of psoriasis. It may well be that persons able to follow a fresh vegetable and fruit diet etc. as he outlines in a sustained fashion would need the herb-juice treatment less often or not at all. It is a great relief to have come across a successful program of action in treating this disease — without drugs! — thanks to Mr. Crosby.

**Mr. Crosby.**

Heart attack July 29, 74 caused from stress, overwork, strain. Went to the doctor. While he was in the hospital his wife brought him lecithin and vitamin E. Started him on cleansing diet, by October upon recommendation of his general practitioner, went to cardiologist, had full physical, x-rays, blood, urine, tred mill confirmed severe heart attack, gave him Fredrickson diet, walking two miles, or four miles ten minutes per mile. Come back in a year. Three months after attack went back to work forty hours a week, walked two miles a day, 2000 units vitamin E, 1000 mg. C, 800 super B. 104 mill lecithin, 2 capsules a day nervine herbs, 2 capsules ginseng, 4 saffron.
Recovery; His doctor stated he was probably in better condition now than he has ever been. He said he knows now when he is tired, to go home and rest, not overdo, to be aware of "what I'm doing and what my body is doing". 180 pounds, 6 feet tall,doctor said lose 6 to 8 pounds, now down to 168.

**My Dear Sister in Christ;**

It was a revelation to enjoy your seminar and it ended all too soon. Thanks for the knowledge, and may God bless all your endeavors and keep you healthy and increase your energy. As I promised, the leper photos, for you to use as you see fit and may they impress the doubters, that seem to multiply. It all began thirty years ago last Christmas. I'll be brief, I saw at one time, eighty lepers at the Lepersario Palo Seco, Panama., and it changed my life. Degradation itself, and having committed no crime, they are prisoners, behind locked doors and gates for the rest of their lives. When I explained it to my wife, she too got the fever and we searched for five and a half years in Europe and finally discovered the formulae on the Island of Rhodes, in the Agean. We were searching for a cure for leprosy only and found among the books, many cures for many things. And recently, Lupus was added to the cures, praise God! Your book may bring hope and release from bondage and I wish you would mention that the same formula will help and in fact, cure Lupus. It is far worse than Cancer, as you know and they are helpless and we must help them. Please get in touch with Mrs. Cook, she too, is carrying the torch hopefully, that one day they will all be cured and accept our Lord and Saviour. Bless you in your work.

Sincerely,
Kirkley

**Irene Hotten**

A young girl (28) complaining of menstrual problems and a severe bladder infection. Periods accompanied by severe cramps, flow was black and stringy, bladder infection odorous and passage painful. Reading of eyes showed spots in bladder and ovarian areas. Colon so black and distended it frightened me, huge mass in area of Sigmoid flexure. Young lady claimed she was not constipated, that she had regular movements every four or five days!!! She had never had either a colonic or an enema and had to be instructed in the use of the enema bag. A complete change of diet-juices for one week then fresh vegetables and fruits, yogurt and exercise. One month later period started black and changed to red on third day with lessening of cramping. Still cannot get any normal activity of bowel. Second month brought mobility of bowel on its own without enemas. Periods normal red and no cramping. When asked what had happened to the bladder infection she replied, "Oh, I forgot all about it". Eyes showed a cleansing of the entire bowel area but with some residual deverticulas. The relief of the pressure of the huge mass that had been in the sigmoid area releasing the ovary so it could empty following ovulation and allowing the bladder to return to its normal position thereby emptying, entirely cleared up the bladder infection. No drugs, herbs or medicine of any kind were used. Just common sense!!

**Vida Walters**

My physical problems began when I was a very young woman. As I went from one doctor, to another for help it was diagnosed as anything from nervous stomach to sluggish digestive system. I was always led to believe that the problem was not serious and their diet and medication would help me. Sometimes it did for a few months then all the old symptoms would return. As more and more foods seemed to give me distress I was merely asked to delete them from my diet; which was already restricted to mild and bland foods. My condition continued to get worse and there seemed to be no place else to go for help. I was tired much of the time and very discouraged; not knowing where I could find someone who could really understand my problem and give me some help. Eventually a method of correct eating habits, using mostly fruits and vegetables, was explained to me. Within a three month period of time I was feeling the effects of this new diet. The stomach aches I had been used to were gone and I had the vitality to do things that I hadn't enjoyed for years. This was over three and a half years ago and I am still happy with the program and am grateful for those responsible for helping me. My entire family has benefited from my experience and have learned to see the value of good eating habits and we are grateful for the good health that we are enjoying.

**Life was spared**

My life was spared for the second time in two years. On Thursday, October 30, I had a dry tickly throat, I started right away sucking vitamin C and went to a meeting in Taylorsville, down by S.L.C., to hear Stan Malstrom speak. I had a carload that wanted to go down, I had a twenty year old girl from Rock Springs with Multiple Sclerosis, who wanted to go so bad. She was going to school in Salt Lake City, I was to pick her up there and take her to the meeting, which I had taped and is very beautiful.

The next day I was sick and stayed in bed all day, Friday, and until noon Saturday, felt better, got up at that time to go to a meeting I had scheduled in Hillyard I felt was another must. I came right home and went to bed and stayed there till 2:00 Sunday. I had a kitchen to draw for my husband which I went to the home of the Browns and did this, by 5:30 I was really sick again so I told Raymond, my husband I felt I needed to go to the hospital. Upon arriving there, I was asked how long I had been sick, I said this was the third day. The nurse told me not to expect the doctor to come to me then, they only came on the first day of an illness. I tried to explain I thought I had strep throat, I couldn't breath and my right gland was as large as when you had mumps. The nurse looked in my throat, called the doctor, said it was a little red.

My husband talked with the doctor, he said go home and take a couple of aspirin, gargle then swallow it. I was horrified, I knew aspirin would have choked me to death. The nurse said before this to take vinegar and salt and water and gargle which I did. That night I sat up all night. Every time I tried to lay my head down I felt I would strangle to death. I also had been taking Golden Seal and Comfrey and Fenugreek, but by 3:00 I knew that unless that mucus started to drain I was doomed. I read in Natures Medicine Chest No. 2, a part on sore throat with Cayenne pepper in a tea. I at once started to take the tea and right away the mucus started to break up, only in little pieces.

By the next morning I was very very sick. My husband had to go to Salt Lake, so as soon as my son got home from school at noon, I had him drive me to the clinic and the doctor took me right in and said I should go right into the hospital or I could be treated at home with salt water gargle every hour. He gave me a shot of penicillin, I went home and did what he said and was a very sick person. I was spitting up every few seconds but still I could not lay down and breathe. That day my girl friend called and said to take slippery elm in a tea which I did, this helped the mucus flow some. By the time my husband was home at 5:00 I was still unable to breathe only sitting up. He asked if I wanted to be taken to the hospital. I said I felt that I would like the Priesthood to bless me. He went right to the phone and called our home teacher, he picked up our bishop and gave me a beautiful blessing. At this time my girl friend brought me some penny royal, she put some in water in a electric fry pan, set it in the bath tub and I was instructed to open my mouth and breathe it. It relaxed my throat so at 3:00 in the morning I was able to lay down but I couldn't sleep a bit on the couch or anywhere, I finally layed down by my husband, and was able to relax and sleep for one hour. By the next day I felt well on the way to recovery. I remember as a child when I had my throat like this I was ill for at least two or three weeks, this time I was really ill, badly, for only three days, isn't that a beautiful testimony? When you take vitamins and herbs, your body is on more of a condition to throw off disease than when it is run down and not able to heal itself.

**Problems solved with mild foods**

I have for a long time had real hate feelings towards my

husband when he didn't do exactly as I wanted him to. I also had no desire for any intimate contact. These things have completely gone when I remain on mild foods with iodine, licorice and chlorophyll. I would, whenever under stress at all, wake up each morning as if I had a bad head cold coming on. My nose was stuffy and would burn to the point of being quite painful. I would also have burning eyes and the feeling of a partial ear ache. When I stay on the mild foods I don't have this problem any longer. Before I started on the mild food program I had continual backache.

After I was on the mild food and no herbs but licorice for about two weeks, my backache left and only when I go back to eating concentrated foods do I get the backache again.

Previous to going on the mild foods, my feet were so rough that I couldn't put on my nylon stockings. The skin would snag when I went barefoot and rip and bleed. Since being on the mild foods (it took about eight weeks), my feet became as smooth as silk and have remained this way. Many times I have given my children lobelia with cayenne, when sick to the stomach. Whether they vomit or not doesn't seem to matter but in a couple of hours after sleep, they feel completely well.

**Doyle Walters**

I have a testimony of the healing power of herbs, mild foods and vitamins. I had a duodenal ulcer build scar tissue and stop the flow of food out of the stomach. This stoppage was critical and would have required surgery resulting in the removal of 2/3 of the stomach, but with the use of herbs, vegetable juices and vitamin E, I was able to bring this condition under control.

This was not an easy thing for me to do. It required tremendous will power but the results were well worth the effort. I am not forty pounds lighter, my energy is way up and a sick day is almost non-existant.

**Ruth Ruhland — Diary on a healing crisis** — picture page 50

Skinny formula — started May 7th on 000 size, three per day and I went to twelve one day, but mostly six to nine. May 9th my neck and head hurt me so bad I went for an adjustment. I had cramping and went to the bathroom often, about six to ten times. One black stool on the evening of May 10. May 13 had another adjustment, my right side of neck and my right ear were so sore I could hardly touch, my ear started a yellow discharge, rather hard and then on to grains, now July 12 it is about clear. My desire for food has lessened and I'm not hungry. May 14 my neck on the right side is so bad I can't move it. My heart picked up a few extra beats in the morning. Last night (May 13) I went sound asleep and really relaxed. I had three hours of sound sleep, woke at 6:30 but wanted to lay in bed. I just couldn't get going until noon. June 3, I've got a lump on my head right above my spine. It's so sore, I can't lay on my back in bed and can hardly move my neck. The glands are swollen on both sides of it, about one inch away. The desire for food has really gone. I could eat once a day to keep from being so dizzy. June 10, the lump is gone but now I'm having eye trouble, can't see to read even with glasses. My far vision is ok, it's only close. My heart picked up a couple of beats but seems ok. Under both my arms and right legs I have large lumps and puss and blood. This lasted ten days and went away but came open and drained. My bowels are still loose (June 20) and I'm on nine a day. I'm six pounds lighter, and was away for dinner the other night and got ill when I was served a steak. I had to go vomit. I crave juice, so I'm making carrot juice. I can't stand coffee (that's unusual).

July 9th, so sick, so awful sick, have had a headache across the back for last three days, the most awful sore throat, and aching, strange feelings down my legs as I lay in bed. My eyes have gotten better. I passed strings of blood, and fresh blood in the urine.

July 10, puss just came out in the urine on its own. I had a very difficult time urinating until the puss and dark strings came out of the bladder. Others were clear as I put a tampon in to be sure. I really worked to get strings and puss. I believe a tea cup of puss. I believe my bladder is tipped again. I felt like I was going to die. My right side of nose drained puss, and I've lost my voice. The left side started today, July 12, to be clogged. I had 103 temperature. and July 11, at night I soaked my pajamas three times and had to change. The bath water was oily or real different this morning. My urine is clear but I'm hoarse and feel like I have congestion in lungs. I'm drinking grape juice, I'm really bloated, my eye lids are so swollen and I feel terrible. Tomorrow I'll stay in bed. I'm on golden seal, ½ tsp. three times a day with one teaspoon of licorice each time. I just ate a pint of strawberries today, I was so hungry for them. Weight loss eight pounds, even bloated. I feel terrible tonight but this is best I can report.

I'm still on the herbs. My nose is all hard and yellow inside but I use cream on it. I really had no sharp pains, my headache is dull now and the most distress was on right side where the gall bladder came out, I did hurt terrible in there. I found it very difficult to breathe, yesterday especially. I do believe I had a crisis. I had a total jet black stool on the 9th, the fluid with the stool was even black. I feel better and I'm sure Sunday in bed all day will help.

* * *

Ruth told me how much better she felt after this cleansing crisis and she looked better than I have seen her in several years. As you can observe, the healing crisis only lasted two or three days. We can have the good and bad days up until the crisis, while the waste is eliminating on the bad days, a little at a time until we eventually reach the point where the body vitality is sufficient to experience a healing crisis.

* * *

**Problem solved with mild food**

At the time I was studying and taking classes from LaDean, I conversed with my sister about what I was learning. Because of my studies in iridology I began to notice people's eyes and my sister's eyes were particularly interesting to me because she had one or two dark spots in each iris. I hesitated to talk to her about them, not knowing how she would accept it. After a few weeks I observed that one spot became larger and very dark. My sister had complained of back problems and pain which she suspected to be gall bladder. Her doctor, after x-rays and checkup, told her he could offer no help or determination as to what the problem was or what to do with it so to let it go for a time. The spot grew larger, it was located in the right iris beginning in the ascending colon area and spread up into the shoulder area. The pain was in the upper right side of her back, under the shoulder.

After hearing in class about a man with a similar problem and what he had done for it, I suggested to my sister to go on a mild food diet, take psyllium seed powder in water with laxative every night and also take a high enema each day for at least three days. She continued with this treatment for two weeks. She noted the spot in her eye had faded. She told me she thought she had passed a bowel impaction. When I saw her after two weeks I detected no spot left in that area of her eye (iris).

**Hearing Helped**

1. Chiropractic adjustments
2. Special blessing
3. Apple juice three day cleanse
4. Full mild foods, three months
5. Herbs, cayenne, apple cider vinegar, B vitamins, started passing chunks of old mucus. Cleared throat and spit out a large chunk apparently from the eustachian canal judging from its shape, after that, level of hearing went up, old aid defective, used loaner only need on low level to hear. Other benefits; chronic, rotten breath gone, asthma mucus condition gone, hair starting to grow in, teeth start to bother, trip to dentist revealed four impacted teeth, after removal hearing on right side went up noticeably. Age 40, deaf since six months of age, 1963 complete ear test, hearing loss so pronounced, both ears, hearing aid useless. Nerves conducting messages to brain dead. 1965, suffered back injury, series of chiropractic adjustments, following this, hearing came in for few seconds at a time over a period of weeks, very spastic. 1970, hearing test again, hearing level in left ear in the low range, other ear no response. 72-73 went on mild foods, discontinued as too expensive, went on vegetable diet, took LaDeans advanced iridology class, began in earnest, husband went on full program.

**Children helped with herbs**

My eleven month old baby had had a bad cold for over two weeks, but with juices, garlic enemas, and lots of vitamin C, we had been able to stay just ahead of it. One night, in the middle of the night, I was awakened with the kind of croupy cough that makes one jump straight up out of bed. I went in to him and his breathing was really labored. I took him in the kitchen to look at him, to keep from waking my daughter. He was obviously running

a high fever. I was shocked at the way his face was swollen. His eyes and cheeks were puffy. I thought, "Oh no, we'd better get him to the hospital fast." As I left the kitchen to go into my bedroom to wake my husband, the word "Lobelia" came to me. I ignored it, and again I felt, "lobelia". A week earlier, I had asked a good friend what I could do to help my baby, in addition to suggesting garlic enemas, she had brought me a few lobelia capsules. When she gave them to me, I thought "no way". I had read in several books the drastic results that came from using lobelia. When the feeling came to me the third time, I turned and went back into the kitchen. I thought, "It's now or never". I took half a capsule and mixed it with a little honey and put it in my baby's mouth. Neither of us was prepared for what happened. Within a few seconds, he threw up a big mouthful of mucus and phlegm. There was a large beach towel on the back of one of the dining room chairs, my children had been playing with earlier, and I grabbed it to have the baby spit up in. I no more than folded it over, then he did it again. Each time, he would catch his breath, then up would come some more. He began sneezing, his eyes were running and he broke out in heavy perspiration all over. He continued this way for about fifteen minutes. When he got through, he was wet all over. I was shaking like a leaf, but had had a deep calming feeling that I was doing what was right, and that everything would be okay. When he finally quit, I looked at his eyes, to see if he was okay. He smiled at me, and in a few minutes was down running all over the house. After I recovered, I made him some peppermint leaf tea, and rocked him to sleep to make sure he was okay. He didn't cough again that night, and within a few days was completely well.
Another time, my eleven year old daughter came from school complaining of a bad upset stomach. Her face was flushed and feverish and it was obvious she didn't feel well. I headed for the enema bag, and she said, "Not that, give me a lobelia, instead". So I did. Within a few minutes, she threw up, went to bed that night feeling better and got up and went to school the next day. I'm convinced we got the infection while it was still in her stomach and before it had spread to her intestines and on through her body. Needless to say, lobelia has become a much used herb in our home. We agree completely, that it is the "Intelligence" herb. It goes where it is needed and does the job that must be done. Its results are very different each time, depending on what the person using it needs.

**Denice Stokes**

It is not just a figure of speech, when I say, "I'll be eternally grateful." After my second baby was born, my health, daily deteriorated. I had pain in my abdomen, and the pain in my back became worse with each passing day. Doctors said x-rays show nothing in abdomen, and finally concluded it was arthritis in my back. Also, I was diagnosed to have geographic tongue. To make things worse my new baby seemed horribly unhappy. I stayed awake nights listening to his raspy breathing, and checking to see if he had rolled over, because when he laid on an arm or hand it would soon turn black. Doctors said; "Oh he'll grow out of it." My nerves were raw, and I was at a loss. No one was able to help. Doctors acted as if they didn't care. I asked myself many times why I had to resign myself to bad health. My Patriarchal Blessing said, I had a strong, healthy body filled with vigor and vitality. Something went wrong! I asked my Heavenly Father to help me. My burdensome, unhealthy body was making me a crummy wife, mother, friend, and child of God. The Lord began to open doors. And the best thing of all, LaDean's book; *Is Any Sick Among You?"* It seemed to make sense. And as I began her diet, I felt my Heavenly Father's presence. My heart and spirit were alive with enthusiasm, ( a feeling I hadn't had for a long time). Within three days after beginning the diet, my arthritis pain disappeared. I noticed my tongue became normal if I didn't eat meat and salt. I woke at 3:00 in the morning, after I had been on the diet about two months. I was in horrible pain in the abdomen area. I could hardly move. Seven o'clock the pain subsided. I passed a very large group of worms all tangled together and then mucus. My baby (nursing from my improved milk) began to act better within two weeks. Every day he passes old filth and mucus, and sometimes parasites. When he gets an earache or a cold I doctor him with herbs. We are not only saving money, but those unnerving hours spent in the doctor's office. My big treat is to examine my eyes at the end of each day. They are filled with healing white, and are surely turning a heavenly blue color. I feel I've been rewarded ten fold for sticking to the diet, October 5. Had another crisis, October 16, I passed horrible looking stuff. Baby also had crisis (I think), I can feel my arthritis pain moving out!

**Paul and Vee Hull**

Following is a short synopsis of some of our healing experiences; I have suffered from eczema over all of my body since I was in the third grade after having been given sulphur baths and sulphur salve. I was approximately 41 years old when we changed our diet and stopped all chemical medications. After about one year of no chemicals, I took medication for a kidney infection, my eczema broke out in fury. On another occasion I took aspirin and received the same reaction. We decided at this time to find the right herb for anything we needed for healing. I am entirely free from eczema now. I received a third degree burn on my hand and by immediately using Aloe vera gel and vitamin E, I completely healed my hand in two weeks with no scarring. New skin formed under the dead white skin and the damaged skin just thinned out and disappeared. At a later date I fell on the back of a metal chair and broke two ribs. By using comfred and a poultice which I have learned to make, my ribs healed totally in two weeks with no scar ridge in the bone and nor further pain. I didn't miss even one days work and some of my associates said they wouldn't believe they were broken if I hadn't had them x-rayed.

Our daughter had her wisdom teeth pulled and instead of using the anti-biotics and pain killers we used vitamin B complex and calcium for the pain, vitamin C to pull out infection and a solution of golden seal and myrrh as hot as she could take which she held in her mouth to heal the incision and prevent the swelling. She woke up the next morning with no swelling, no bleeding and no pain.

Paul had critical health most of his life. His eye showed severe inherent lesions and a star burst colon. He couldn't exercise as it fatigued him and made his muscles ache. We discovered through iridology that he had Addison's disease and with the information in LaDean's book we have seen improvement in his health over the past two years. Just recently his health has taken a dramatic turn for the better and for the first time in his life he can work long hours without fatigue and without muscle soreness. We have a testimony that what LaDean is teaching is the Word of Wisdom as the Lord would have us live and through following the basic principles we have received those "Hidden treasures of knowledge" regarding our own families health that are promised in the Word of Wisdom. My husband Paul and I have both had serious health problems. When we met, married and combined six children from previous marriages, we also had health problems with our children. We were both interested in health through herbs, nutrition and eliminating preservatives in our diet, so we went at it wholeheartedly. We first took salt out of our diet — gradually so the children did not know the difference. It was not until they visited their grandmother and wondered what was wrong with the cereal that we told them they had not been eating salt. We next switched from sugar to honey and from white bread to sprouted wheat. As we eliminated the preservatives, used more fruits and vegetables, stopped frying, reduced our meat intake and learned to use herbs instead of chemical medications, we found exciting health changes. Our oldest daughter had chronic infection which kept her in and out of school about six weeks a year. She was given high doses of anti-biotics to "burn" the infection out of her system and gama globulin shots to help with the infection. The hours of hospital testing gave no answer as to the cause or any solution to ending the treatment. After Paul and I were married and changed her diet, her health straightened out, she ended her sleepless nights, kept good attendance in school and lost the white pallor to her skin. Her hair which she was told would have to stay short, grew long and beautiful and her stooped posture straightened up. Our younger son, plagued since infancy with constant tonsilitis, regained his health on our new diet and hasn't had a reoccurrence in five years. At the time we started our diet change, we searched for information on herbs but found it very limited in our area in 1970. We were told of LaDean Griffin's class about two and one half years ago. We both attended and received information that changed our lives and our health. I have a testimony that we were led to LaDean and that she teaches by the spirit. In LaDean's class we learned of the mild food diet and I immediately started using it. Within two weeks I had a healing crisis in the colon. I had a forceful cleanse which lasted all night but by morning I felt as though atmospheric pressure had been taken from me. There was no effort in doing anything. Paul and I observed the mild food diet faithfully for over three months. At the end of July we went to a family reunion and my 25th class reunion-both the same day, and ate what was served. I ended up flat on my back, sicker than I remember being for years with a cold, the first sinus infection I have ever had and lung congestion. I learned that my system could no longer tolerate the preservatives in food or the bad combination. Paul and I signed up for LaDean's advanced iridology class and at this time Paul developed a method of photographing the eye for LaDean's classes. Since that time we have continued to do eye photography, learn iridology and increase our knowledge of herbs and healing.

**Karen Biddulph**

I can't remember a time in my life when I wasn't plagued by bad health. Even as a youth I didn't seem to have the health and vitality that my friends had. In my high school years I had the same problems. After I was married and began to have my family, my health steadily deteriorated until about five years ago, I was plagued constantly with heavy depressions and illness. The depressions were so bad that I constantly feared death and I feared any disease. Any ache or pain was played up in my mind to be a terminal illness, and I would constantly live in fear of death. I had times of depression that I just didn't seem to be able to overcome. Along with this I had a physical problem and I checked with the doctor many times and was only given tranquilizers to keep myself from worrying about it. It got so bad that I couldn't drink water without my stomach hurting. I couldn't eat any food. I had constant pain in my spine and I had headaches that would never stop and, of course, the depression came right along. After the birth of my fifth child, I didn't have the energy to even take care of my family. I'd watch all day as the children would run around in their pajamas. I wasn't able to take care of my home, and I wasn't able to take care of my church responsibilities. Sometimes I'd wonder if my husband and my family might not be better off if I were dead. I was told at this time by my doctor that the best thing I could do for my depression was to see a psychiatrist, and I was almost ready to but I sought help from some church leaders and I was told that psychiatry wasn't the answer to my problem, and that if I'd be faithful my Heavenly Father would come to my aid and help me out of this problem I was so deeply involved in. So, I began fasting with my husband and praying that my Heavenly Father might see fit to bless me. I have a firm conviction that all laws which we gain in this life are predicated upon obedience to the principles on which this law is based, and I prayed with all my heart and my soul that my Heavenly Father might see fit to give me this law. In the meantime, I studied the program theory. I studied all the books I could read on nutrition and health. I tried the vitamin supplements. I tried the strong protein drinks and the high protein diet until at last I felt my head was just like a big pumpkin askew on my shoulder. I couldn't even hold my head up at times. My scalp hurt so badly that it even hurt to lie in bed and, of course, the depression became worse and worse, till I felt that I was at the bottom of a black pit and couldn't even see the top. After about a year of constantly praying and asking my Heavenly Father to help me, during this time I asked if I should go to a clinic or follow the orthodox medical route and always I felt inspired that I should not do this. I have a very strong admiration for the medical profession and some of the things which they accomplish and I am grateful they are here for things which we can depend on them for, but in my own particular situation I felt this was not the answer. After a year of this searching, I was told about a series of lectures that were being given not far from our home by LaDean Griffin. A friend of mine said that she would pick me up and take me down to hear the lectures if I would go. Before I went I stepped into my bedroom and knelt and asked if the Lord could please in some way help me that I wouldn't fear disease. The fear of disease seemed to be the cause of a lot of my depression. As I sat listening to Mrs. Griffin's lecture, the tears rolled down my cheeks because one of the first things she said was, after we learned things, she would teach us we would no longer have to fear disease, and I felt this was a direct answer to my prayer. I listened to the lectures completely, and I had a firm conviction that they were true. When we studied the part on iridology, I was afraid because after looking in my eyes I could see that there were many parts of my body that were deeply diseased and that it would take a lot of dedication and faith to live the program as I should in order to get well. I came home and explained everything that I'd learned to my husband, and he immediately knew that it was true, so our entire family went on the program. We changed our diet to that of mild food. We began taking the herbs that we needed, and we began to take the vitamin supplements, and all those things that we had been taught, to the best of our ability. The results were almost remarkable. We had a little boy who was eighteen months old who had been on gamma globulin since he was six months old. When he was four days old, we took him into the doctor with a terrible sore throat inflammation and from then until he was eight months old it was a constant round of gamma globulin, anti-biotics, and hospitalization for different diseases. After just three weeks on the mild food diet, he was able to stop all of his gamma globulin shots which we had been giving to him every six days. He hasn't had a shot now for two years. He's a healthy little boy and he's growing stronger and stronger. Knowing what I do about iridology, looking into his eyes, it's a miracle that we found this program for him, as well as for myself. During the first year of living on the program, I had a baby — a beautiful little boy, and he's the picture of health, he's strong. His resistance to disease seems to be almost unbelievable. He's an intelligent little baby, and we're very grateful to have him. The rest of our family, also, is showing great improvement in their health. I must admit that during those two years I've had many bad days, and only knowing that this program is true and the testimony which my Heavenly Father has seen fit to give me, has sustained me through some of the bad days. I went several weeks many times without having a good day. I'm finding now after two years that the good days are outweighing the bad days and I have a strength and a vitality which I didn't think I would ever have. Another thing that I'm grateful for is that I've been able to get well here with my family. As I have recovered from the illness which was in my body, I've been able to take care of my family and do the things which I could. Another thing which I'd like to mention that Mrs. Griffin taught to us which is as important as anything that she taught us, was that it's possible to be able to listen to the Spirit and have the Spirit guide and direct us in all the things that we do. We've had many experiences, my husband and I, of having a sick child and after administering to him, knowing instinctively what to do and being guided by the Spirit what to do for our children and having them respond almost miraculously to the things which we were told to do. We have taken our children through colds and fevers, ear infection, and many different, minor illnesses, and we've been able to take care of them without having to take any anti-biotics, and we've known which herbs and which vitamins and what kind of therapy we should give them. The blessing that has been most important to us is knowing that our Heavenly Father lives and that He's willing to give us His Spirit to guide us in the things which we should do for our family. We've learned iridology. We've been privileged to take the advance classes in iridology, and it is remarkable and further testimony to us that this is true because everything as we looked into our eyes corresponded to places that we had been having trouble in our bodies and after two years on the program, it's a miracle to see how our problem areas in our eyes are interlaced with the healing lines which Mrs. Griffin promised would come if we were faithful to the program. It is our testimony as a family these things are true. They have brought nothing but goodness to us. They have brought nothing but peace and harmony in our home and feelings which we have not had in our home until this time. There have been many among our associates who have looked down upon us and ridiculed us and called us fanatics for the things which we have done but we feel that we've been following the course which our Heavenly Father would have us take in healing our bodies and becoming more spiritually in tune with Him and living the way that He would have us live. The colds, flu and the illnesses which our family suffered have been the result of living on the program. They are healing crises rather than actual disease crises. Our family has been strict, and I've seen rewards come from living on the program the way we have done.

**Kent Biddulph**

My wife Karen has already explained the circumstances in which we became acquainted with the mild food and herbal remedies program as taught by LaDean Griffin. This program has been an answer to our prayers for improved health. Even though it may not be the ultimate in what our Heavenly Father intends, it is a considerable and sizable step above the conventional diet of the general American populace which is accepted by the medical profession to be correct. This viewpoint I cannot accept. I have not taken all of LaDean's classes as my wife has, but I do know this program is correct and follows all scriptural references I have been able to find and study on the subject. In my own way I have tested this program on myself and found very definitely that if I follow the program as outlines, my health, mental attitude and spiritual growth is greatly increased. If I do not follow the program, I do not enjoy the benefits. My experiences after two years of living and eating as taught in the mild foods program and seeking our Heavenly Father's guidance reaffirms to me that this program has indeed been an answer to our prayers for our family. I can give my personal testimony of its worth and the truthfulness of the program. I know it can help anyone who seeks after truth and good health.

**Marilyn Parkinson (Mr. and Mrs. Bob Parkinson)**

We are thankful for the opportunity to leave our witness,

along with many others, in this book. We have watched our eyes change through following the mild food diet and through use of the herbs. We would encourage anyone who has a desire to be self-sustaining and independent to study iridology and make use of it in caring for their families. In this day of rising medical costs and strikes and shortages, it is encouraging to know that there are simple and practical and inexpensive answers to many of our problems. Thanks to LaDean Griffin for the effort that she has put forth to bring these truths to us.

**For Tanyet McKenzie**

She was seven months old at the time the program was starting. I was nursing her at this time, and when I went on the mild foods. And soon after this she began a real cleanse. She had been an extremely nervous baby. She wouldn't sleep at night, she cried continually, someone had to tend her all the time, and yet she seemed to be healthy. During this time I managed to get some of the nervine herbs down her and tried to take all the things away from her that would cause this nervousness. This, in itself, didn't seem to make much difference. But as I increased my licorice root intake, this helped some and then I put her on alfalfa in the form of liquid chlorophyll, at the time I did this, within two days she was like another baby. She quit crying all the time, she would sit down and play, and before this she had been so tense and so nervous that when we tried to give her an enema or anything of this kind she would cry and be so upset that we simply couldn't do this, she would just close off. There was no way to get water into her. But after this and she calmed down, then we could give her enemas and we started moving things from her, and we could tell my looking at her eyes, even though we hadn't had a good picture before, that they began to change. I thought that they were almost a black-blue color, but as this began to move the toxic waste, she started to feel better, her eyes are gradually getting to be a lighter blue, and she continues to pass an awful lot of waste, far more than the amount of food she eats. At this time she is sleeping through the night, she is happy for the most part and she has color back in her face, her eyes don't have the dark look that was around them, and I contribute it more than anything to the mild foods and liquid chlorophyll.

**Cheri Munns**

My daughter May was four months old and I was having a hard time nursing her, as with my first baby. Up until six weeks after I gave birth, nursing worked beautifully but then I would start my period and nursing became a problem. I hadn't started May on any solids or bottles, so that wasn't causing my problem. I did everything my breast feeding book said but with no results. Well I was becoming a nervous wreck and my daughter started sucking her thumb to fall asleep since my nursing her every half hour around the clock didn't satisfy her. I was convinced giving her cereal wasn't the answer so I poured my heart out to my Heavenly Father to know what the answer was. Why were many young mothers nursing their babies so easily while others like myself were having such a struggle? I got so desperate that I threatened the Lord that if I wasn't able to nurse my daughter the way I knew nature had intended I wouldn't have anymore children. This was quite a threat for me because I had always wanted a dozen children or more if I possibly could. Well, my prayers were answered when the manager of the apartment complex we lived in brought iridology into our conversation one day and then told me about the mild food diet and the reasons behind it. I then read the book, *Is Any Sick Among You*, and after living on the mild food diet for two weeks, I was able to satisfy my baby with more milk and an easier let down. After another month on the diet I stopped having periods, the way nature intended to have birth control. My girl is now 1½ years old and I am still nursing her beautifully. I thank my Father in Heaven every day for the wonderful blessing of health this new way of eating has brought to my entire family with the added help of herbs.

**Healing Crisis**

After being on the mild food diet for about six months and thinking I would never have a big healing crisis, it happened! At first I thought that I had caught the flu. I ran to the bathroom all morning and then it really hit, like being in labor. I had severe abdominal cramps, except they wouldn't let up. My diarrhea turned pink which I assumed was watered down blood with mucus in it. After awhile no more blood only straight white mucus which turned yellow, green brown and then to my utter horror greenish black which came out in globs. Before this crisis I had trouble with foul smelling gas whenever I ate dried figs, dates, raisins or avocadoes. After my crisis, they never bothered me again to eat them. I feel that the crisis I experienced cleaned my whole bowel out of all that morbid mucus waste that had probably been there most of my life! It was well worth the pain to be that much further on the road to perfect health.

**Mariphyllis Foreman**

I had melanoma cancer and the doctor wanted me to have surgery. I decided first to try herbs. I took comfrey and blue violet leaves, put them in the blender with honey and lemon, I drank ten to twelve glasses a day for two weeks, and then went three to four months on raw food. After that length of time, I went to the Cancer Clinic in Mexico and they found no trace of cancer.

**Rhona S. Hanks**

I am a member of the Church of Jesus Christ of Latter-day Saints. About three and one half years ago, I became very interested in studying the 89th Section of the Doctrine and Covenants where the Lord gives His people some guidelines for good health. I cam to the conclusion that there were many things contained in this section that were far above my level of understanding.

A well-meaning Sunday School teacher of mine when I was a child, told our class that herbs were vegetables. I didn't question that statement until I was thirty years old. I made it a matter of fasting and praying to my Father in Heaven for answers to my questions about good health procedures and to increase my knowledge of the Word of Wisdom. Within a few weeks, I was inspired to study other books concerning the uses of herbs in healing the body and more correct foods to feed my family. From this information, my husband and I decided to try some of the things.

Our youngest boy who is eight years old, has been sickly since birth with one thing or another. For the past two years, we have tried to feed him as many fresh fruits and vegetables as possible, as well as fresh raw vegetable juices. We have cut down on his intake of sugar, refined flour, and meat. We have added vitamins, minerals, and herbs to his daily diet. Last year in the second grade, he only missed two and one half days of school because of illness. We have been so thrilled, that is a fantastic achievement for him.

From LaDean Griffin's lectures on health, I learned about the purge for the gall bladder. My husband had been having what our doctor suspected was trouble with his gall bladder for several years. I had had my diseased gall bladder removed at the age of twenty two, and my husband knew well the difficulty and expense of that kind of operation. He decided to try the purge. He passed so many stones the next day that we stopped trying to count them, they were of varying sizes and in many shades of green. We estimated that there were over two cupfuls of stones. Since that time, which has been three years, he has not experienced any more related symptoms.

For over three years, I've been telling myself that the mild food diet was for sick people. If I ever got really sick, I'd try it. I finally quit rationalizing, bolstered up my courage, and went on the mild food diet three weeks ago. Granted this is too short of a time period to test it, but this much I can say, not only do I feel good and am losing weight in the process, I'm beginning to feel a part of the sensation that comes with self-mastery.

Whether it is putting golden seal on a cut finger or treating the flu with mild food and herbs, our family has been truly blessed with this knowledge. It is my testimony that the Lord answered my prayers and directed me to these truths.

**Tumor passed on mild food diet**

We are blessed to have many children, but it was disappointing to my husband and I when I lost a baby about two months along. I then heard about and read LaDean's book and started on a mild food diet. A few months later I was pregnant again. I began taking herbs when a bleeding problem developed. Just past two months I lost this baby at home. I felt so bad, but still I had a peaceful feeling. A few minutes later I was impressed to know that I wasn't through. I began to bear down and passed a long tumor. What a blessing to see health returning to my body.

**Teresa Clingler**

Thank you for the love and understanding that you have for everyone. You're so beautiful, and I love you so very much. I hope with all my heart that I can learn to be as loving and beautifully spiritual as you are. I knew from the first time that I met you that my prayers had been answered. I'm so thankful that I am a daughter of my Father in Heaven and for some of the understandings I have of

Him. I plan to learn a great deal more so that I may be able to raise my children to walk uprightly before our Father in Heaven. The only thing that is strongly planted in my mind is to teach them to know while they are just little to take, and live with the herbs and mild foods so that soon, some day, they may be able to walk uprightly before our Father in Heaven. It's so hard to imagine how wonderful it would be and I want my children to be part of this. I came to know of the herbs and mild foods when my baby was born. He was born October 22, 1974. When they put this baby in my arms, I knew there was something wrong. He was perfectly physically, but I didn't really pay too much attention to the warnings I had then because I didn't know too much about the mild food and herbs, but after two weeks had gone by, I knew I had to do something. I took him to my doctor, and I told him I needed some help. There was something wrong. I told the doctor he acted as though he was in great pain.

He slept beautifully for a week and from that day on it seems as if he was going downhill. He couldn't sleep. He had a hard time breathing. He was so white and so limp. I have a little girl three years old, and I know how a newborn is supposed to respond and grow. The doctor sent me home with two big pockets of prescriptions and changed the baby's milk to a formula — soybean milk. I gave him this prescription and milk for a month. There was no change. Now he could hardly lift his head at all. It seemed when I'd pray to my Father in Heaven I couldn't pray without soaking the sheets with tears. I prayed so hard that He would answer my prayers.

After that, I took my baby to the Idaho Falls doctor. He put him in the Idaho Falls Hospital for a week. They x-rayed every organ in his body and took lab tests. There was still no change. The doctor still didn't know what was wrong, so he put him in my arms, and he said, "I guess it's just because he's so used to the attention he gets," so he told me to take him home and put him in a room and let him cry.

I took him home and I gave him a mother's blessing and when Dee got home that night, we got everyone together and gave him another blessing and that night he had no trouble sleeping, and I was able to sleep myself.

The next day my mother had talked to a lady about mild foods and herbs and my prayers have been answered. I was ready to do anything because I knew my baby was dying. We had his eyes read, (they were a natural blue,) they were such a dark brown that the parasite sunburst and the deterioration had turned them so dark you could hardly see the pupil. We put him on fruits and mild herbs and through fasting and prayer I have a live and happy baby. After five weeks of herbs and fruits, he started expelling tumors, things that you would never dream would come out of a pure, perfect, newborn baby. I knew my Father in Heaven was close and my baby's little body just beamed. I knew I was doing the right thing.

When I held this baby I felt so close to my Father in Heaven. I haven't been to a doctor since December. He still has a long way to go, but he is alive and well. His eyes are starting to turn a light blue. I have never in all my life seen anything so wonderful, so spiritual as the herbs and mild foods. It has brought our family so close to our Father in heaven.

**Verlyn Ritchason**

I have had a thyroid (goiter) problem for nine years. I've tried ½ grain of thyroid medicine, about eight and one half years ago. I used it for about one week and it made me very nervous, so I quit using it. About three years later a health food owner told me about kelp. So I took about six small kelp tablets for a week — then I was very nervous so I quit taking that. Then the thyroid enlarged and started hurting. Then I took an herb combination of irish moss, kelp, parsley and capsicum — I took one the first day-it made my thyroid hurt-none the seond day-one the third day-a little hurting-none the fourth day-one the fifth day-no hurting. I have been taking one or two every day since March. The goiter has diminished to half the size and no hurting. I feel great!

**Georgia Stephenson**

My name is Georgia Stephenson. I live in Orem, Utah. Last August, 1974, I attended a class given by Mrs. Griffin on her book, *Is Any Sick Among You?*. I have recently over the last three years been interested in health and felt there were many things about health that the average person didn't know.

I went to her class after reading her book and was somewhat skeptical about some of her ideas concerning health and illness. I had never been what you really could consider a sick or sickly person. I've never had anything really seriously wrong with me, but I have had very troublesome acne since about twelve years old. I have spent a lot of money on dermatologists, etc., with not much help from them. I have also been a good thirty pounds overweight for about seven years and have tried every diet that has ever been thought up.

After attending Mrs. Griffin's class I decided to try mild food and some herbs. I have always been a nervous person and unable to stand much stress. I started right on mild food without any problem at all. I continued to be very active with housework, yard and had a garden, children and a husband to take care of, not to mention a lot of church duties. To my surprise I felt great and was starting to lose weight. On any high protein or regular reducing diet I would lose weight and then after two weeks I couldn't stand it any more, but with mild food it wasn't really hard not to cheat and want candy, etc.

Then, besides losing weight, my face cleared up and stayed that way except for two weeks when I was off of mild food at Christmas time, but as soon as I went on it again it cleared up. Also, within a few days of starting mild foods, I noticed my neck glands and throat felt a little sore. Not really hurting, just noticeably sore. The only illness I had had in the last six months was a sore throat. After Christmas when I started back on mild food after about a week, my nose started running a little. Then I had a real bad headache and realized I should be taking high doses of vitamin C so the mucus would dissolve and be able to get out easier. I did this and the next day my nose really started to run and mucus came out of my lungs also. These are the areas you can see in my eyes and eye slides that are weak and they are cleaning out and healing. Also, the parasite lines are going away in my eyes since taking black walnut.

Besides clearing up my skin, it's such a good texture and I have rosy cheeks for the first time since I was a little girl. Probably the best thing mild food and herbs have done for me, is something I didn't even realize until about a month ago, I don't have headaches anymore, I have always been headache prone. Let me get under a little pressure or have to hurry to get things done and I'll get a headache. It runs in my family. Anyway, one day I realized I hadn't had a headache in months. I am First Counselor in the Primary. I was in charge of the children's presentation at Sacrament meeting. Anyone who has had anything to do with these will know what a responsibility and challenge they are. I thought, "Well, this will really see what sort of shape I'm in." I went through several practices and the program, as worried and under stress as usual, but

I didn't have headaches. This is possibly my strongest testimony that mild food and herbs work for me. There is no possible way I could have gone through these practices and that program without several headaches and lots of aspirin. I have always been that way before.

**Carol Oster Miller**

In giving my testimony of this program, I do so with much thankfulness. I have known for many years that we are, to the most part, what we eat. Through the years of soft drinks, french fries, and hamburgers, somewhere in my conscience, something warned me to take wiser care of my body, but I didn't listen until I needed to when much damage had already been done. Strangely enough, these problems didn't reveal themselves until I was married and expecting my first baby when I had a nervous breakdown. This, of course, was brought on mainly by stress and a different way of living. My doctor treated me with tranquilizers and from that point on all things — all problems medically that I had — were attributed to nerves by the doctors. This stress started many other problems until after my fourth baby when I found I had osteoarthritis and ankylosing spondylitis, a hardening of the spine. I was told I would eventually need both hips replaced. There was no hope. Just years of suffering to look forward to. I felt absolutely defeated. Then through a good friend I heard of a program of mild foods and herbs. Everything my friend told me seemed to ring logically true. I wanted to know more. LaDean Griffin, the lady my friend had learned the program from, was coming to give lectures, so I attended.

At first I couldn't believe what I was hearing. It was too simple. I prayed to know if it was true, and I received a beautiful, spiritual testimony of it. I live on the mild foods diet. I am taking herbs for my particular problems, and I personally believe they are helping me regain my health and strength. I know I am healing. It won't happen overnight, but I do know I will again have good health.

I've learned through personal experiences of friends that this program is true. I have seen it work. Through taking the herb, black walnut, for instance, I can walk without pain, and before I was in constant pain — severe pain in my hips. Walking at times was a thing I had to mentally do to put one foot in front of the other and force myself to move. With black walnut I can walk and not feel this pain. This program has brought me closer to my Father in Heaven. I have joy and a greater faith in tomorrow. I can weather the storms easier and rear my children in a quieter, more peaceful home.

Simple truths have always been a stumbling block for people through the ages. I believe this program, simple as it is, to be part of the hidden treasures spoken of in the Latter-day Saint Word of Wisdom. God does provide if we are willing to search out the truth. For every problem there is an answer somewhere written, if we will search it out. For this I am truly grateful.

**Dorothy Cook**

When I started taking herbs I knew I would eventually start feeling better. What I didn't know was the full impact herbs would play in my own health and that of my entire family. I get so excited when I recall the change that came over me in just a few days of taking licorice root and a combination of gotu kola, ginseng and capsicum herbs. Absolutely a miracle! It is so marvelous to wake up refreshed, full of energy, more tolerant of people and events, and a deep sense of well-being. Needless to say, I had no trouble getting my husband and children on herbs after they witnessed the change in me, and I'm happy to report the changes in them have been just as rewarding.

My husband and I are with each other constantly, as we own our own business and actually enjoy all this togetherness. Our children still tease and quarrel with each other, but with a much kinder tone of voice which in turn has stopped fights from erupting. In fact, our lives have reached the point where if one of us uses a tone of voice other than the nice tone we have so quickly gotten used to, the comment made in response is, "Take your herbs!" We are all taking other herbs and food supplements now for certain health problems and weaknesses, but there is no doubt in my mind that licorice root, gotu kola, ginseng and capsicum have made us feel so much better that we have the courage and desire to continue making the changes necessary in order to have more perfect health. Thanks be to God for His wisdom in providing herbs for His children and for answering my prayers for better health and happiness for my family.

## Overcomes tumors with herbs

I was sick most of my life, due to weakness, so when I found out about the herbs and suplements, I felt it was a blessing and an answer to my prayers. I've always wanted to help people and this was one way I could. Since I've been on the herbs, I've really noticed a change in my health. I had rheumatic fever as a child, three to five years old, and started developing arthritis, (rheumatoid) about twelve years of age. I had a weakness in my legs and my back (due to falls). Now, my hearts almost healed up, the weaknesses in my legs is almost gone and arthritis is diminishing.

I also chewed gum quite a bit from ten to sixteen years of age; I was very nervous and had two ulcers during high school years. Half the time I chewed gum, I swallowed it. About two months ago I passed some gum. Last month I went on Doctor Jensen's garlic fast, and passed plastic. Since I have been on the herbs I have lost a fibroadenoma tumor. Doctors had decided it was a genetic defect and during removal of my first tumor, they saw my tissue was nodular. My second tumor caused them to conclude I would have more, but lucky for me, not cancerous tumors. I shrank that tumor with massive vitamin E supplements, then when I started taking herbs, I cut back and developed my third, which the combination CS shrank in two days.

My health has been improved in other ways too, and I believe I owe it all, first to the Lord and second and third to my loving parents, and my friend for being with me through this, and I know the best is yet to come.

I also lost approximately ten unwanted pounds in six months with mild food diet, lecithin, and a combination of chickweed, licorice, saffron, gotu kola, mandrake, fennel, echinacea, and black walnut herbs, that helped my glands function properly.

## Goiter Disappears

I live in Bakersfield, Calif. I came into the knowledge of herbs some nine months ago. I have however been acquainted with the herbs for about twenty years and didn't realize it. In 1961 a surgeon told me my youngest daughter would have to have a goiter removed from her throat before she choked on it. I have baby pictures — she was born with it — because of my diet — no eggs, greens, etc. On comfrey, kelp, and garlic tablets, I had her well in one year. I took her back to the doctor, he measured it, and it was gone and has never returned. I myself had a corn between my little toe for twenty years, it is now gone. According to iridology, my eyes were brown, now they are hazel, no more gold colors in my brain area on the right side. I have a deformed rib, an injury, we found in my eye I never knew I had, that spot is almost gone. Also where I was kicked by a horse about ten years ago (in the right ovaries), also showed up in my eye with two spots, one is gone and the other going. I have had spotting in ovulation ever since the injury, now with the black cohosh, I do not spot any more at all. I have had all the miserable things that women have to put up with at that time of the month, and I am happy to say that I do not even know when the time is arriving (which can also be a problem.)

My husband has not been able to sleep on his back since I married him because of a pain in his right shoulder and headaches. We gave him the gallbladder treatment and he passed 75 (or more) stones and some the size of my thumb. He sleeps like a baby now, no more pain in the shoulder. I had sinus also when I first got onto the herbs I had more headaches than ever, and then I would drain and drain. Now I never have headaches and that is really great.

**Earache and Pneumonia**

A friend of mine had a five year old boy who had a terrible earache. He'd had it for over a week without success from medication the doctor had given him. Out of desperation, my friend contacted me, and I told her about a garlic enema and vitamin C. She proceeded to give him 1000 mg. of vitamin C every hour and painted his throat with iodine and glycerin every fifteen minutes. Within two hours the swelling in his throat, which caused his earache, was gone and the pain was too. Within another hour, the child wanted to eat again. We gave him fruit for another twenty-four hours, with the vitamin C. By then the infection was gone and never returned. I have a dughter who has had pneumonia several times. It has never been an infectious type and had no fever with it, but still had all the other symptoms. The doctors never know what to do, as she does not respond to anti-biotics. When I learned about the mild food diet, I put her on it. She has never had pneumonia again, however, if she gets too many concentrated foods, going off of the diet, she immediately starts with congestion, so she has to be very careful not to eat them too often or she could end up with any of the lung diseases. As for my own self, I had a period of time when I didn't feel good because of an adrenal problem. I acquired arthritis from this, so I now take licorice root, I feel better but I have a long way to go. My body was so weak that building with vitamin therapy was very essential to me. I now have about three weeks, pain free and ten bad days. I look forward to more good all the time.

**Parts of a letter to Reverend Crosby**

**Dear Reverend Crosby:**

I have just completed ten days of taking your Herb Tea Formula. I cannot believe the results! This simple remedy of your herb formula accomplished what thirty-five specialists (including fifteen at Mayo Clinic) were unable to do, and who could not even come close to it in the seven years I have been plagued with Systemic Lupus Enythematosus. The results from your seven day fast, with juices and yogurt, was almost immediate. So much so, that I have not had an injection of gamma globulin for the last ten days. I needed them every three or four days, using 45 CC's in June and July. The only problem I had was with the yogurt, as I am highly allergic to all dairy products.

Thank God I am now free of all headaches, which I had constantly. My illness began with a fever that lasted six weeks, around 101 to 102, and I am surprised that the drugs did not kill me. My hands were all covered with blisters for two years, and felt as though they were on fire. I would put them in ice cold water, to relieve the burning. It was always in the back of my mind to go to an herbalist, and now, you are my salvation. I have had so many injections, I am running out of skin.

The dear Lord heard my prayers and came to my aid when I received your formula. I still have a long way to go, but every day brings surprises and I am sure now that the formula will bring me the cure I have sought for so long.

**Vanda Just**

I have always had problems — health problems — all my life. Always infections. Gamma globulin has always been low, and my children have inherited these problems. Several places in my body were deformed from it. A little over a year ago a specialist told me that within the year I would not be pushing my own wheelchair.

There was much pain with it, many problems, lots of medicine, gold shots, steroids, cortisones, on and on and on. But when the big expense came, it was with lupus. The arthritis had turned inward, and it acts the same as rheumatoid arthritis does, only it eats out the lining of your vital organs.

The doctors told me I was lucky last November because it was in the beginning stages, but it is terminal. I had been in so much pain. My skin was gray, my heart was bad, my eyes had sunken-dark circles around them. It affected the family because everyone suffered. I had gotten so bad and filled with so much pain that one night in February, after my husband and I had gone to bed, I felt my heart stop. I felt my breathing stop, and my spirit came right out of my body. I can be up a dozen times with my children during the night and my husband never knows it, but something woke him up that night. He jumped and he grabbed for me and he called me because he knew something was drastically wrong. I couldn't hear him through my ears. I heard him through my spirit. It sounded as if it were a hundred miles away. The pain came with pushing my spirit back into my body. I felt my breathing start. I felt my heart start beating. I finally said to him, "I'm not going to make it." And he said, "You will now. You've got a pulse."

The next morning when we got up, I asked him if he had remembered this, and he said, "I know so much more than you. I know how close you came. I don't think you'll ever know." We decided to fast and pray because we knew healing was an answer. I had prayed for many years for healing of arthritis, but one night before February I was filled with the Holy Spirit and I felt the power of him in my hands. I touched the places that I was crippled and drew it out, but God did not seem to heal me of the lupus. There had to be a reason.

I still have arthritis, but I don't have the purple places, and it's very much controlled. But getting back to that time. We were talking, and the phone rang. A woman who owns a Health Food Store in Blackfoot called me. There had been a woman who walked into her store, she was from Logan, Utah, and she was a friend and neighbor of LaDean Griffin. She saw the book on the counter and made a statement that her friend was going to be giving lectures in Rexburg, Idaho, and then this friend told her about me and she said that LaDean Griffin had a friend that had lupus and that she had been learning about it and was getting good results, so please have me come. This friend called me and I knew instantly that this was the answer that God had given me. That was the answer to my prayer. I went up to Rexburg. I took the lectures and I quit all medications. I was on $250 worth of drugs a month, and they weren't helping me at all. They were just covering up the problem and there was still so much pain and agony and misery. In two months I have improved so greatly that our family attorney asked me for help. I referred him to LaDean's book and have helped many people this way. But what also brought it home was the fact that I've got a four-year old child. All of his life he has had health problems like all the other children I've had. He's been on several gamma globulin shots a month. He's had anti-biotic shots, many of them a month. Bottles of anti-biotics. Constant medicine, and he would lay there like a vegetable with no will to even get up and play, and I started him on liquid vitamin and minerals, high dosages of vitamin C, and that child has not had a shot or medication or anything in three months. He's a strong, healthy, normal child. I know his body has to clear out a lot of toxic waste, but he's well on the road to recovery. My thirteen-year old several weeks ago had a very severe asthma attack, caused from being around horses. His fingernails and lips turned purple. His eyes rolled in the back of his head because of lack of oxygen. It's very frightening to use herbs, just herbs, when you've always used the hospital and oxygen and you wonder, "Am I taking this child's life into my hands?" "What if I don't do right?" It's frightening. For thirty-six hours we worked over that child, and he came out with flying colors. I've never seen him improve so quickly as he did on herbs and vitamin C. A year ago someone witnessed to me about herbs and I thought to myself, "Each to his own," and I didn't want to jeopardize my life by taking myself off of drugs and going onto herbs, and yet, really, it was the other way around. I feel that the Bible proves that herbs are for the use of man for medicine, and I really feel this is a lesson that God wanted me to learn otherwise He would have healed me of the lupus and taken all the arthritis away instead of just the crippling. I feel it's going to be the answer and I really feel that if doctors would become more aware and use preventive medicines, such as herbs, in their practices that we wouldn't have so many problems. Many doctors are turning to this, and it is proving to be very beneficial.

**Shirley Yancey**

I took the classes from LaDean Griffin about two years ago and have lived close to the laws of health, mild food and herbs to heal the body. Since I was a very nervous, depressed individual, living this program of health and reading the scriptures pertaining to health convinced me, this is the way our Heavenly Father chose for His children to live because of its simplicity, not cutting into the body, and side effects of the drugs.

Taking these classes from LaDean Griffin has shown me this law. When we live this, it gives us a freedom — freedom from the terror of dreaded disease. We have the assurance that something can be done and we, with the helpful prayers to our Father in Heaven who knows all laws, can free us from this bondage. As a mother, this has paid its dividends many times. Reading and studying, going to LaDean's classes and using herbs has brought our family through many illnesses and pain, but also as a mother it brings many burdens and heartaches. It's very difficult to try to convince members of a family to change their eating habits when we use the science of iridology and see that they have sick bodies but they are not willing to give up the traditional all-American hamburger, the hot dog and soda pop, to change from their meat, bread, sugar, and tacos to the use of fruits and vegetables. This is where the burden comes when those whom you love have used drugs for many years but their bodies are deteriorating with pain and cannot see the truthfulness of this law. These are the burdens that you have to bear. You also meet people who have been searching for this law who have gone the route of surgery and drugs and finally found this new hope and are surprised by its simplicity. I must say that this has brought me much closer to my Heavenly Father. In order that we know the truthfulness of all things and feel His Spirit must guide us whenever we change our lives for improvement we need His help. I've called on Him many times and felt of His direction. I feel fortunate that over two years ago I went to LaDean Griffin's classes and could partake of her many, many years of study and research, that my health can be changed. I feel fortunate also for this blessing to me that I can help others and this law will bless their lives.

**Shirley Rawlings**

I went to LaDean's classes for three days, the next day my father came. He had had a real critical accident in a saw, with his hand, and he almost lost his little finger on his right hand. It had cut him in several places. When he came to my home, it was just heavily invaded with infection, and it was brilliant, scarlet red. Immediately I saw the infection, and so I decided that it needed to be soaked and that there must be an herb that would draw this out, so I mixed a paste of lobelia with vitamin E oil. I put the paste on his finger, and then we put just a small bandage around it so that the air would still circulate. It hadn't been on his hand very long and he looked at me and he said, "Boy, that is really potent salve that you put on my finger. I can feel it begin to draw." Well, we kept this on for three days, and we kept soaking it three times a day and put the lobelia on three times a day. By the next Monday we took the lobelia off and just started putting golden seal and myrrh mixed with vitamin E oil on it. About a week-and-a-half later, it was healed and starting to close. The infection is all gone, and we're just keeping vitamin E oil on it to keep it from scarring.

My little boy had a wart on his finger, and since I knew that the lobelia would pull things, I decided to put it on to see if I could pull the wart. After I'd put it on the first night, well, the next morning you could see the little roots start to pull up and you could see the roots come off and within three days the wart came off.

**Peggy Burnshaw**

Lavon: "When did you first hear about the mild food and the herbs?"

Peggy: "When Sister Griffin came last August."

Lavon: "Did you stay on the mild food after that?"

Peggy: "Yes, in fact I stayed on it for twenty-one days without stopping. I didn't snitch once. I was so proud of myself. I felt better than I've ever felt. In fact, I'd just had my baby. She was about three months old. I had a terrible pregnancy. My nose was red the whole time. My nose and my face were red and my back was always breaking out and things like that. I couldn't understand that, and I felt terrible, always tired. I was eating liver until it came out my ears. After I started on the diet, about three days afterwards, my nose completely cleared up, and went to the normal color of my face. I was eating a lot of avocados — it was that time of the year, my skin turned really nice. In fact, I felt really good. I must not have been very weak because I didn't have many bad days at all. In fact, sometimes I felt like I could just fly. I just felt so good."

Lavon: "Did you take any herbs at all?"

Peggy: "No, when we first heard it, I was nursing the baby and you know, all the propaganda that goes around, what you need to nurse your baby, this and that, and I was just so concerned about losing my milk, and so when I went on mild foods, I figured I'd better not go on anything stronger, so I didn't go on CS or anything like that for a long time and then it was about a month before I got going on licorice. You know, it takes all this stuff awhile to sink in until you can understand it, and so we didn't do very much of that, but with my baby I took a lot of vitamin C when her nose was running and that really helped her and I gave her several enemas, and that would clear her fever up just really fast. My husband went hunting and they ate meat and other things after he'd been on mild foods for a long time and he was really sick. Anyway, his eyes are bad, and when he got sick, everyone said, and he said, "Oh, I know it's the meat. That is what's done it." Mel's kind of crazy anyway. They got Aspen Bark and made a tea of it, and it cleared everything right up. They all said it's because it would taste so horrible it would cure anything. I've been taking licorice now really alot and the CS, and my baby's eyes are just really changing, and they were really horrible for a little baby. It scared me to death, but not anymore, you know. She'd have probably gotten really sick very young. Her eyes were almost a brown-red color, and now I think they're going to be gray and there is really a lot that has cleared out. My husband's eyes have changed and my eyes are also changing.

I haven't stayed on the diet, like at Christmas time, we blew it, and ever since then it's been like a week on and three days off and a week on and a week off. But always our diets have improved just enormously. And so, from our eyes, there's a difference. I have taken a lot of herbal pumpkin and lately a lot of iodine. Lately, really strong on things, so I'm sort of anxious to tell."

Lavon: "Are your eyes fading off?"

Peggy: "Yes, I'm anxious to see my next pictures. My husband's eyes have changed a lot, and he's taken the black walnut. We've taken a lot of it. Black walnut and CS. Lately I think my vitality is up because I have a white tongue, so as long as my tongue's white, I don't worry."

In order to eliminate we have to have vitality, and we know when the tongue whitens, the elimination is beginning to happen.

**Melva Spivak**

I first started on this program in August, and at this time I started on the mild food diet and I went on it up until November. I gained forty pounds on the mild food diet and that was a real shock to me. But what happened was, I filled with fluid, then my kidneys stopped functioning. In December I saw your mother, LaDean. She taught me a whole new idea — mild foods and herbs. I set myself up on a program. The kidneys started functioning again, but I was still retaining water, and I know that I am a hyperglycemic. I've learned this, and I have heart trouble. Before this particular time I had been on heart medication for a good five years. I went off my heart medication and all drugs completely. I wasn't taking any after that at all. And I had somewhere from six to eight spells with my heart, and I wasn't familiar with the hawthorn berries. The only thing I really was aware of was how well cayenne worked, so I got on cayenne and started taking three capsules a day. Whenever I'd have a spell, I'd take two teaspoons in water, and it worked just like nitroglycerin and stopped the spell instantly. This was quite an experience.

Lynn: "Did you stay on the mild food diet strictly?"

Melva: "I did up until about December when I gained all this weight, then I got to panicking and thinking and reading and I studied on hyperglycemia, and I thought, boy, I'm not doing the right thing, and I got to panicking because I was getting so big. So I thought I'd better add a little protein back into my diet, and this is what I started doing. I started back on protein. Well, first I started with taking yeast. I found that eating meat, after that length of time, I got sick, and it bothered me, so I started taking some yeast. In about the middle of January, I had a cleanse, a really big cleanse, and I passed a worm, we decided, that was six inches long. We looked at it, and we decided — we didn't know what kind it was — but we knew that it was of the worm family. You could see it.

So I stayed on this protein, and I started losing a little bit of the fluid, and now I've started back on the diet again. I've lost ten pounds since I came on the diet. But the thing I'm not sure about is, I don't know whether I have Addison's disease or not. I might be in the further state, but I don't know. I've been taking four capsules of licorice a day, and I take vitamin E. I also take a combination of hawthorn berries and capsicum, and I've noticed — we talked about it in class today — that if you have bumps on your thumps in your heart, and I've noticed since the time that I went on this that mine has flattened out and I no longer have these bumps, so I know that the heart has improved. I no longer have the heart spells that I was having.

I can see a definite change in my eyes. I've noticed that the colon has cleaned up considerably. My nerve rings were really bad. They have really lightened and phased, a great deal. The parasite lines have phased away, and there is a definite change in the thyroid. I know that my thyroid glands are still bad, because I can feel them. They hurt at times, and I can see it in my eyes, but it's starting to phase off, too. Possibly, when I reach that certain point, I expect to lose some of this fluid I'm still retaining. I think it's because all the glands are out of balance and aren't working together. I might add, that I decided that I went into the cleanse too quick, because I went mostly on fruit, and a person that's been sick for years — in fact, for ten years I've been really chronically sick. My body was not strong enough to go into this cleanse, and so it threw me in so fast when I started on fruit that I decided this is what stopped the functioning of the kidneys, and I just plain got sick. The kidneys had stopped for a good eight hours, and I was administered to and my name was turned into the temple at this particular time. My kidneys had stopped in the morning, and it was midnight that night that they started functioning again. They've never quit completely since, but they still are not working completely right. One other thing I might add that I have been doing for kidneys. I've been drinking two cups of sagebrush tea, and this has really been helping to keep my kidneys moving.

**Lorna Stratinger**

My husband and my little boy and I have been on the mild food diet for about six months. At first we went on a lot of potatoes because we were really hungry, but we never felt so good in all our lives. And at first we were really, really nervous because I am sure we were cleansing and I have always been a nervous person. In fact, I was on tranquilizers in high school. I'd lay in bed and shake and throw up my food. I couldn't sleep, and things were really bad. Now I sleep like a rock and take lots of vitamin B and calcium and licorice root. I feel like I learned something new in the last classes because I quit taking licorice root and now I find that I'm hyperglycemic, and I thought it was hypo, with the symptoms. Now I take golden seal-one golden seal to every two licorice in the morning, and I do it again at night. Then I take saffron to utilize my fruit, and I can see that that's really helped me because I haven't been able to utilize my fruit before. My complexion has cleared up. I feel so good, and I hope I don't lose any more weight. I feel good, and I have energy. I feel like I want to get up in the morning instead of laying in bed, and that's just occurred in the last two weeks.

My husband does a thousand push-ups a day and he boxes and his eyes are full of white. He's so strict he never goes off the mild food diet. He knows that — we feel that this is a celestial law and we'll be living this way some day, so we want to try to clean our bodies up and learn about herbs so that we can help others who may not be able to learn about this. The time's coming when they'll need help and when they're sick and can't get doctors.

Lavon: "Have you noticed a change in your eyes?"

Lorna: "Yes, I have a really, really bad colon, and it takes lots of jigs and jags and it has black, destroyed tissue there and destroyed tissue in my ovary area, that's why I'm not pregnant. That's the reason that I'm taking the formula, and I'm taking it twice a day and I can see white in quite a few areas of my eyes now.

### Geraldine Cook

I have had severe migraine headaches for some thirty-five years of my life. I have tried doctors, medication. The winter of 1975 I spent in Mesa, Arizona, and I went to Phoenix to three headache specialists. They took the brain wave. They took the brain scan, and they put me in the hospital for two days, and they didn't find anything; so, when I came home my sister told me about this program. I got a letter from a doctor at Stanford University saying that they wanted me there on the sixth of May for one more test, and this was after LaDean's first class. I was so impressed, that I wrote a letter and I told the doctor, "I'm tired of drugs, I'm tired of getting nowhere. I have found the answer." So immediately I went to the health store and got the herbs that she told us about, CS, and I'm on the lower bowel, but there's another for headache.

Lavon: "A combination of valerian root, wild lettuce and capsicum?"

Geraldine: "Yes, and no headaches. They're going," so I told him, "I'm going to do it naturally this time because at last I had found the answer, so I wouldn't be there." So, maybe they'll look into it and see that this is the only way. I went on mild food. I had my husband on mild foods. He was treated in Arizona, also, for one month, four days a week, for Emphysema.

Lavon: "How's he doing?"

Geraldine: "Well, I've got him on this program, off the drugs, onto the lung herb and also the hawthorn berry for his heart, and we're on the juices. We went over and bought a juicer, so it's changed both of our lives, but I thought I was doomed to have these migraine headaches, and I'm engaged in the field of genealogical work, and I wanted so bad to get back to it and now I feel the Lord has helped me to get here to this class."

**Gloria G. Lowry**

In 1960 in the late fall I became aware that there was something very wrong. I went to my doctor, and I said, "There is something very, very wrong with me, and I'm not really sure what it is," and he gave me a rather perfunctory examination, and said, "Oh, all you need is for spring to come."

I didn't really believe it, but I accepted it. In April when we returned from Conference I went back to him, after having had a very dreadful experience with extreme nervousness, tearfulness, and feelings of hostility, and I went back to him and he said, "You have a double goiter." I had a hyperactive thyroid gland and my eyes were already bulging. It took him until July to build me up to the point where they dared to perform surgery.

During this time I lost a great deal of weight, became very, very nervous and began to experience and manifest symptoms of neurosis, and so after the surgery he said he would release me in the care of a psychiatrist, and he did. This psychiatrist was just opening his practice and was trying to make a name for himself and I suppose, really, I became a kind of guinea pig.

They began to treat me in the tranditional way with psychotherapy and drugs. I alternated between depression and rebellion, and as I would begin to rebel against the treatment he would raise the level of medication, and I stayed with this man for about nine years, during which time we paid him $20,000.00 in drugs and medical costs, and he told me that I was schizophrenic and continued to adjust the medication to try to help me maintain some level of stability because my body could not handle these things. It was a very difficult situation, and I became depressed and suicidal on two or three occasions and he put me into a private mental hospital and treated me with shock treatments and medication. It was a very difficult experience. I've had twenty-one shock treatments. Eventually, we went to Washington, D.C., on a Sabbatical leave. He told me that I would always have to take therapy and that I would have to have periodic shock treatments throughout the rest of my life. When we went to Washington, D.C., I realized that I did not need to see this man every week or every two weeks as he had led me to believe, and although I was functioning on a relatively low efficiency level, I was functioning. While I was in Washington, I had periods of dizziness and a great craving for sugar and candy, and I tried all these years to live the kind of life I had been used to living — creative and busy and active, and it was very, very difficult under that kind of medication, as I slept a good deal of the time. My children, when they were very young, thought that I was a witch. So, while I was in Washington I craved sugar, I remember on one occasion I went to a candy store and bought a two-pound box of peanut brittle and ate it in one sitting, and I continued to indulge myself in sugar because I was so weak, and I thought I was taking a high-energy boost.

On Mother's Day that year I took the children out to dinner. My husband was out west on business. We had a lovely dinner and a dessert, and then we went back to Sacrament meeting. Following Sacrament meeting, I became quite dizzy and faint-feeling, and I said to the girls, "We must leave." I made it just barely out of the chapel into the foyer and passed out, so they put me into the hospital, and the doctor who had been treating me — I found a doctor when we went to Washington to supervise the medication which I was taking — was called when I went into the hospital, and he didn't even speak to me. I was dizzy. He said, "What are you doing in here?", and I said, "Well, I passed out in church," and that was the last word he spoke. He did not examine me. He did not touch me. He did not do anything at all, just wrote on his pad of paper and then released me. And I was extremely dizzy and very, very ill.

My neighbor was extremely upset. I was having hearing problems. I couldn't hear the doorbell ring, and there was noise in my ears, and so she caller her doctor and arranged for a specialist to see me and he said, "They should never have released you from the hospital," and they put me back in. They did a blood sugar test. An internist came and did a blood sugar test, and he said, "Your body doesn't seem to handle carbohydrate well," and I said, "Does that mean I have diabetes?" And he said, "No, you are not a true hypoglycemic, either, but, I would suggest that when you return home you go to an internist and have this checked out." Well, I thought, you know, that if I didn't have diabetes and I couldn't handle carbohydrate and I was gaining weight so rapidly that I was just going to get big and fat if I didn't stop eating candy, but I couldn't seem to.

During the years under the psychotherapy the psychiatrist felt that it would be a tragedy if I were to have another child, and he insisted on putting me on a birth control pill, and I gained fifty pounds in a year. I insisted on going off the pill. When you're under this kind of medication, you really do not have control. You think and you have desires, but you are not able to express them and certainly not able to assert yourself against a powerful personality, and this was the situation in which I found myself. So, when we returned to Michigan I went back to this psychiatrist, and I said, "I want to discontinue treatment." I said, "I will take the medication, but I want to discontinue treatment." He said, "No, you must always have someone to supervise," and I said, "Well, all I really need is a chemist," and he smiled and said, "Well, I'll be your chemist." Well, I didn't want him to be my chemist. I didn't want anything further to do with him, but I believed that I would always have to take thorazine, and so I agreed then to continue the treatments, so that I would have another prescription written. He wrote them only for the period of time between visits, and this was going to be a once-a-month checkup. So he gave me a prescription for thorazine that would last a month. Then, when I got home I called and told him that I would discontinue treatment, and he sent a real scare letter to my husband.

Now, over all this time he was treating me, I was on various levels of medication. For a period of years I was on 350 to 375 milligrams of thorazine a day, and then when I became depressed on that, he then gave me a reduced dosage and an anti-depressant, so in effect I was on an upper and a downer, as the young people today say. Then, I came west with this one month of medication. I had great fear in my heart because I really believed I had to have it, and I didn't have a doctor. I talked to my cousin, who is a medical doctor and he said, "There's nothing schizoid about you. There never has been." And he said, "If you really want a psychiatrist, find someone else. Don't take that kind of a diagnosis from any one man." So, I went back home, and I wanted desperately to get off the medication, so I went to a general practitioner, whom I had not met. And he said, "I do not understand this drug, and I would be reluctant to take you off it." He recommended another psychiatrist in town. Now, he took me off the drug immediately. He didn't work me off from it at all, he just said, "Stop taking it." At that time I must have been taking, perhaps 175 milligrams of thorazine and I don't know what the other one was, the upper. But, I had tremendous withdrawal symptoms. He continued to check me and after three weeks, he said to me, "I wish that we had more people in the world with your maturity and with your stability, it's interesting to me that the other psychiatrist was not able to reprogram your mind," and I said, "Well, there were two things going for me, I have a very strong spirit and the Lord was on my side."

And the Lord continued to be on my side, but I was very frightened because I was alone and, as I would go to a doctor for help, they would want to know, what is it that's wrong, and as soon as they would find out that I'd been under psychiatric care, they would label me as neurotic, and if they couldn't determine what was wrong with me, it had to be in my mind, and it didn't really matter what the complaint was, it had to be in my mind. This attitude carried over to other members of my family. It was a very frustrating experience. So, I tried to live with it. I was very, very weak. I had periods of extreme weakness and faintness. For about a year I tried to take care of my home and my responsibilities. When the weakness would become too great I would lie on the floor till I could get up and do a few more things, and then I'd have to lie down again. But I continued to fight to maintain some normal life.

Through all this my wonderful, wonderful husband was a tower of strength. Extremely frustrated. He had sought the best medical help he knew. He simply did not know what to do. He did all that he could, having faith that he was doing all he could do and trusting those who were treating me. But the Lord was really watching over us and all these years my name was in the temple and many people prayed for me and encouraged me by helping me to continue to struggle. A friend then called me. I had become ill again after I had gotten off the thorazine and away from this second psychiatrist. He saw no need to continue treatment. And I was just so very, very weak, and I felt as though I had the flu. I went into the doctor, and he tested me. He gave me a blood test, and he said, "You appear to have a virus. I'd like you to go home and take aspirin and go to bed," which I did for two weeks, and I didn't feel any better, so I went back, and he did another blood test, and he said, "This is extremely interesting. Your white count is exactly where it was." He said, "These tests were done by two different technicians, and it's extremely unusual. I would like you to go home and take aspirin and go to bed." So, I did, but you can't really stay in bed when you have a family and you have a life to live, and I began to try to function. Then, I was visiting a friend, and I passed out. We were practicing a musical number we were going to do in church when I passed out. They took me to an osteopathic hospital, and they did many, many, tests. I was there for two weeks, and, as I now realize, I must have been going out — in and out of an insulin shock because I was not able to eat, and so I was fasting. Well, now that I'm on mild foods, I also realize that there were many poisons already out into my system and moving, and they didn't understand it, and I certainly didn't understand it, they would give me additional medication, and it was really a very hellious time.

During that period of time, I had a woman come into my room to be a roommate who had been in great pain and had not been able to sleep for a week, and she was such a brave little thing, and didn't complain at all. I couldn't bear to see her in that kind of agony and I just asked the Lord to bless her that she would be able to rest. I continued to pray and continued to watch her. I raised up on my elbow and stayed on my elbow, leaning up, as long as I possibly could, concentrating on her and praying that the Lord would grant her some rest. She dropped off to sleep and remained sleeping till I was too weak to continue and dropped off myself. She said, "That was almost a miracle, I haven't slept for such a long time." I then gained a testimony that the Lord can heal and the thing that we are taught, that faith can heal was a valid thing, and it moved beyond faith, but to an absolute knowledge that when we ask, having faith, we will be given the answer. I asked, having faith, for myself and I didn't seem to have the answer that instantaneously, and it was sobering and it was a little frustrating. I decided that for some reason the Lord didn't want to reach down and touch me and make me well, that I needed to find solutions to the problem. And I continued in this manner.

I spent two years — two summers — during the past three or four years in Salt Lake City and Provo undergoing many, many, tests, under the care of my dear cousin who loves me dearly and would move heaven or hell to help me, and I realize that. They really and truly could not pin-point exactly what was going on. I have a border-line hypoglycemia. The manifestation was too dramatic, and they really felt that much of this was in my mind. At one point, after I had been two or three wekks in the hospital, I was released, I was in great pain, and they didn't know what was wrong. I became very, very depressed. I had been taking a mild sedative because I was not able to sleep, and he felt I needed rest and apparently it built up in my body.

I got up one morning at Mother's house, and I said, "I'm going for a walk," and she said, "All right, but you're weak. You've been in the hospital for a couple of weeks, so be careful. Don't go too far." So, I said, "I won't." I had an ice pick in my hand and I walked as far as I could walk in a numb, disorganized fashion, as far as I could walk down into a corn field and fell and lay there trying to get the courage to stab myself. But they had missed me and were praying for me and the Lord was watching over me, and I did not do this horrible thing which, of course, I did not want to do and would not even have considered had it not been for the drugs in my system. And I came to, after several hours of sleeping and waking and sleeping and waking in this corn field, and dragged myself to the road and a young man picked me up and put me in the car and took me to the hospital. I was deeply depressed and I refused any kind of medication, and I refused any kind of food and, of course, they can't give it to you if you refuse it now. In years past they could.

Paul became deeply frustrated. He didn't know what to do, and he said, "If we can't get her to eat or to accept help, I'll have no alternative but to put her in the psychiatric ward at the University. The poor little darling has given up hope." My husband called. He, of course, was with the children in Michigan, and he called and he tried to encourage me, and he began to give me a blessing and to pray for me over the telephone, and I hung up. And I attempted suicide that night. I had a glass of water, and I was going to break it and they could sense that and they kept a nurse in the room all evening and I don't really know what happened, but something did, and I changed my mind and permitted them to give me medication and they brought me out of it. But, unfortunately, it was another anti-depressant and it only compounded the problem. It was a very, very difficult struggle to continue to strive, to continue to stay faithful, to continue to have hope, to continue to pray, and to continue to try to function on a meaningful level, but I was able to and I'm very grateful.

The following year I made the acquaintance of a man who uses hypno-therapy, and he, through this technique, helped me to understand my feelings toward the psychiatrist and the understanding that I was not a total loss, and he planted again within my heart real hope and faith. This man, unfortunately, has been led astray. At that point in time he had been led astray by the religions of the East, and many of the things that he told me I countered through my own deep faith and would not accept even in hypnosis. Eventually, I came to the point where I could no longer use hypnosis, and I feel that the Lord blessed me there. He permitted me to have enough of the experience to free myself so that I could move on, and then He removed me from danger. I have a great testimony in the love and compassion and patience of the Lord in spite of the fact that we are so stupid and so stubborn.

I began to have increasing weak spells, falling into a gray-out, not a total black-out, but a gray-out experience where I could hear things going on about me, but I could not respond, and this continued for about a year, and I never knew when I left home whether I would get back or not, whether I would end up in the hospital or just what would happen, and we lived in a kind of a low anxiety. But you can't really just give up and lie on the bed. I guess you can, but I'm not that kind of person, and so early last spring I was shopping, and I felt ill. The doctor had told me to carry candy with me, and when I felt ill eat candy or sugar. It was as though a switch had flicked in my brain somehow in my head. It was almost a physical thing, and then I would sink in varying degrees of rapidity into this gray-out, and I ate the sugar. I was in a large discount store and I tried to make my way to the front, and I just barely got there, then I passed out and they put me in the hospital. I was there for a week and they did tests. There were two internists who came in, new doctors who did not know that I had a "psychiatric problem." Two internists came in to treat me, one of whom believed that I had a real problem and the other thought I was craving attention. Little did he know, because I had all the love and attention any human being could want, and many rich blessings of which I was fully aware. I was in the hospital for a week, and during that time I would go into the insulin shock and they really and truly did not understand what was happening. At one point, when I went into the shock, they had been treating me with orange juice and sugar and I told the nurse — I felt this happen and I said, "I must have some juice and sugar," and she said, "All right," but she didn't come. She was busy with other things, and she didn't come, and when she came I was semi-conscious. So, she attempted to feed it to me with a syringe orally and I nearly choked, I suppose. It was rather ghastly. So, they called for a blood sugar test. Well, I had had sufficient orange juice and sugar that the level in the test was in a reasonable low range of normal, because I was only semi-conscious. I heard the two technicians say, "She's just pulling a bluff. This is not a real comatose situation."

As I came out of it — and I do tend to come out of it, not rapidly and not to a very high level — but the orange juice and sugar apparently had some effect — and I came out of it and remembered and realized what had been said, and I called the nurse and I asked to see the head nurse. They came and I said, "I have been through hell. Please do not put me back in." And the head nurse apologized and said that conversation ought never to have taken place." I said, "But, it did, and I do not want it on my record. I have been through hell. Please don't put me back in."

After a period of time I was released. I was home for just a week again to the store and this time I had my daughter with me because I was afraid to go alone. It takes awhile to get up your courage again.My daughter was thirteen, and I felt this happen to me, and I said, 'I'm going to be ill, Amanda. Go call Daddy." We happened to be standing by a candy counter, and I remembered that the doctor said that chocolate candy would probably pull me out of it more rapidly than sugar cubes. So, I reached over and picked up a package of chocolate bars, but could not think how to open it. I could not remember. I could not function enough to open it. A couple came by, I remember them vaguely, an older couple, and he said to me, "Are you ill?" And I said, "Yes." He said, "Do you have a sugar problem?" And I said, "Yes." And he unwrapped the candy bar, but by then I was too weak to chew it, and they took me into the back room and waited for my husband. They were not able to revive me, so they put me in an ambulance and took me to the hospital. My husband, of course, followed in the car after having taken the children home. During that interval, when I was in the emergency room, I began to have severe tremors and the doctor in charge thought that I was having a granmal attack and he screamed that he wanted dilantin, and they gave me a heavy dosage of dilantin and, of course, I was no epileptic and so it did not have any positive effect. After they gave me the dilantin and it didn't bring me out of this tremor and out of the problem I was having, they decided to keep me and do some tests and, of course, I had already had every test in the book about sixty times, but his doctor didn't know it; so, I went upstairs, and because it seemed to be a brain problem, they put me into the ward where they have people who have had strokes, and things like this. They continued to give me dilantin, and I was allergic to it. I broke out in a rash, and they thought I had measles, but instead of discontinuing the dilantin because the tremors were continuing to come, the seizures, they gave me benedryl to correct the allergic reaction, but I'm not able to handle antihistimine drugs and it only compounded it and I became anxious and then depressed and upset and so they gave me a tranquilizer to counter that reaction and after three weeks time and under these heavy medications it was an extremely difficult situation, and I became very frightened. They finally decided they could do no more, that I would be as well off at home, and I was released.

I decided that I had had all I could cope with, that I did not know what else to do. I had put my faith in the doctors. I had done diligently all that they had asked me to do, and I had continued to pray for guidance. But when I was released from the hospital in this dreadful state, I made a covenant with the Lord that if He would help me to get well I would not put another chemical into my body, and I have honored that, and it has been very difficult to be in pain and to have these situations occur and not to have medicine as I had been led to believe I needed to have. I've found that I've gotten along just as well, if not better, without it. I continued to have these blacking-out spells. Then I went to the Word of Wisdom and decided that I would learn to live the way the Lord wanted me to, because I had long since realized that it's what you put into your body that creates or destroys good health.

While I was in the hospital with the dilantin and all the other horrors, a young friend from our ward came to me and said, "Sister Lowry, you've helped me and I have a book that I think might be helpful to you," and it was Sister Griffin's book. Now, I was under heavy medication and I flipped through it and I read such things as herbs and things and I thought, "Oh, dear, here's another kookie health food book," and I said, "I appreciate your bringing it, but I really have not been much of a health food person." Oh, she said, "That's fine," and she left, and then I thought, "How ungracious of me to have been, not even to look at it," so I called her and I said, "I apologize. I really would like to look at the book," and so she ordered one for me. As I read it again, I thought, "This is a lot of garbage." I guess I just wasn't humble enough.

That was a year ago. Over this past year, I have read the Word of Wisdom and prayed a great deal. I have read everything I can find on nutrition, and many new studies, of course, have been done concerning the addition of bulk and the effectiveness of vitamin C in dissolving cholesterol. I read John A. Widtsoe's Word of Wisdom, and I began to feel that I was on the right track, that I really needed the right kind of food.

It began about a year-and-a-half ago, and at that time I had gone out to eat at a restaurant that weekend and about three days later I became very ill with something that seemed like salmonella only far worse. I had sort of a lockjaw where I couldn't open my mouth and some shaking chills and diarrhea. Anyway, I was very, very ill for about four days. Finally, I talked to a doctor on the phone and he prescribed low motil and tetracycline, which at that time I took. I didn't get better, I just kept with the shaking chills and the diarrhea. In fact, even the American Medical Association has come out and said low motil should never be used, and it turned out later that I had paratyphoid fever which you would never use a strong anti-diarrheal for because it just backs it up in your system and that's what happened. It went all through my body, and into my lymph system and everything.

Until that time I had always felt pretty well all my life. The only thing that I can remember happening was about five years previous I developed sort of a coated geographic tongue that never disappeared. I was sick all the time, yet I didn't feel sick, and I'd asked various doctors about it, and they would say, "Oh, it's nothing, nothing, it doesn't mean anything," and so I just kind of passed it off. Well, then, with this disease or whatever it was, I couldn't get out of bed, except to go to the bathroom, and I just never got better. I just stayed profoundly weak — the most profound weakness — and I'd shake practically all night, and I'd just pile on the blankets. I ran a fever and it would never go away. I had it for five months with no help. I'd go to various doctors and they'd do all these tests. Then finally I wound up in the hospital, and they ran all these tests and didn't find anything, and so, finally I made an appointment with a specialist, one of the doctors did. He said, "It acts sort of like parasites, but we can't find them." Then this doctor sent me to the University of Utah Medical Center and the gastrointestinal specialist, because this doctor in Provo thought that it must have something to do with that, because I had diarrhea and everything. Then he checked me and he said that he strongly suspected that it was regional enteritis which was too bad, that it was incurable, and that the only way to test it was to open me up and go in and cut me open and find out, because there's no way to find it except by the symptoms, and my symptoms were classic symptoms of regional enteritis. He didn't recommend cutting me open because there was nothing they could do for it anyway, and so — the symptoms are the fever and profound weakness. The only way they try to take care of it is, they keep taking out part of the intestine and the colon to kind of halt it a little bit to try to give a remission to the disease, and it doesn't work very well at all. It's not a cure.

Just about two weeks before this I was in desperation and I got in the car one day and I decided, 'I've got to find the answer,'' and, of course, I had prayed diligently all this time and I can remember saying, "If I don't care where the answer comes from; I'm willing to do anything, you know, to get the answer," and so I got in the car, my husband said he would keep the children. I just drove, and I went over to a clinic — it was some baby doctor clinic. It wasn't even my baby doctor, and I went in there and a nurse took me by the hand. She said, "Come here and talk to this doctor about your problem." And I went in there and he told me, "You know I have seen cases of salmonella in children that they couldn't even find the parasite or anything," but he knew it was there, whether their tests found anything or not. He said he's sure because he could give the drugs that they give for it and alleviate the problem. So then he took me by the hand and said, "Just remember, the medical profession does not know everything." That hit me just like a light bulb, and so I got in the car and I started driving. It was the strangest thing. And someone had told me before that there was an herb shop near my home and I went to this herb shop and walked in the door and just said. "I have been ill and I just don't know really what to do." She sold me some herbs, I think it was red clover blossom combination and some for nerves because by then I had never been in such a state. I was just a nervous wreck because I couldn't take care of my family. I had none of my family members here to help me, so I just didn't know what I was going to do. These herbs were teas, some herb teas. I started using them and I didn't really see any results. For one thing I was eating exactly the way I had before. I wasn't doing anything about diets. Nothing about diet, so I continued the same. I grew up in Virginia where you eat a lot of pork and you eat a lot of meat and you eat a lot of greasy foods.

My parents came to help me. My father and mother were there for two weeks, and I didn't get any better. I had started on these teas and my father and mother said, "This just can't go on." I had already had a blessing at the first of the disease, but my dad said, "We'll have to give you a blessing," so my father-in-law and my father blessed me, and the next morning I went upstairs. I never passed anything strange in my stool before, but that morning and that whole day I passed nothing but white mucus after that blessing and so I knew that there was an answer. Even though I kind of realized I wasn't going to be cured, you know, overnight or I didn't have the faith, I don't know what but I knew, there was something in there. You know, the doctors kept saying there's nothing in there, nothing in there, and when I asked a doctor in Salt Lake about it he said, "Well, we don't know what causes it. We don't know what it is. We don't have any idea, and there's no way to get rid of it." But Bill and I saw all this mucus coming out. You don't really pass all that much with the disease but you do pass mucus with it occasionally, but not this great. In fact, that whole day that was all I passed. That was sort of an answer, so I knew there was an answer.

I went down to another health food store and there was a sign up that said "Iridology", and of course, I was intrigued by that. I decided to take this class. I could hardly get out of bed to go to LaDean's class.

Lavon: "Then you heard about her lectures?"

Gloria: "Yes. I read it on the sign. Iridology class, by LaDean. I went to her class. I could see the doctor was wrong, that I could probably cure this incurable disease."

Lavon: "What was your feeling when you heard LaDean?"

Gloria: "Oh, well I came home and I can remember telling my husband that night — so weak, "You know, there's very few times in my life I've really felt that the person that was talking to me was inspired and I was getting just about nothing but the truth, you know, and this occurred tonight."

Lavon: "This is what you prayed for?"

Gloria: "Oh, yes, I know it's an answer to my prayer, that I was led, step by step to her. So, I knew it was the answer. I knew I had a long way to go, but I was going to try it. Then I started the mild food diet.

Lavon: "You bought her book, I guess?"

Gloria: "I bought her book and studied it cover to cover and started taking the CS and the herbal pumpkin and then I started the strangest thing, too. I took the fig formula and that was the first herb I really took. I hadn't started the other one, and I couldn't believe it. Because of her book and changing the diet — you see, the other herbs didn't do much, when I was taking those teas I didn't really notice them, but when I started the diet and took herbs — Wow! My nose started immediately cleaning. Well, in fact the first day I ate nothing but fruit. That was probably a mistake because I was so weak."

Lavon: "Cleansing fast?"

Gloria: "Yes, well, I haven't really been that sick. What happened was that my nose just drained and my eyes and my neck hurt right where my lymph glands are and I wasn't expecting this, and it took me another class of LaDean's to realize — to know what to expect. A couple of days passed and then it would hurt in my lymph glands and my arms."

Lavon: "You told me you had lumps under your arms?"

Gloria: "Well, it kind of swelled up a little bit and then it would hurt and, of course, some of these things are sort of frightening because I'm new at this, but I still realized that I had been led to this answer. I tried to keep my cool. So, then I stayed on it exactly three months. In fact I just couldn't believe it. I got really, really sick. It was the first time I'd gotten really sick. But the interesting thing, the doctor told me this fever would not go away, that I'd just keep it. After being on this diet for two weeks the fever was gone from this disease, so that was something. In three months I had this big crisis, which I just couldn't believe. I passed nothing but mucus and ropes of mucus — black, dark brown and black, that color, you know, poisonous. It wasn't the white that I'd had before, that I'd had that one time when I had the blessing. I guess — I don't know what the black means compared to the white. Maybe it's digging deeper, but anyway, I was very sick. I was only sick for two-and-a-half days. It was just amazing. I couldn't believe I could be that sick for that short period of time and then get better."

Lavon: "You must have passed that poison out in that short period of time?"

Gloria: "Yes, and then it left and then I was better than I was before.

I went back to Virginia and stayed on the diet, I'm sure there were some things that were in the foods, I know that Mother would occasionally slip butter into the peas and things like that, but I stayed on the fruits and vegetables and then four months later, I had another healing crisis. I can remember, at this fourth month. I hadn't eaten for two-and-a-half days, and yet I went in and passed this thing at least a foot-and-a-half long of rope of just straw. It was just hard. What it was well, it reminded me of when you're clearing a drain and it's gotten all stopped up and it's been stopped up for a week or something — a month — and you know how it's kind of slimy and black all up in the drain. I passed so much that I'm sure it was at least the size of my body, of stuff, and I can't believe it could be in there. But, see that is the classic regional enteritis disease. They've shot dye through people that have had it, and the lymph is totally congested and that shows in my eye — the lymphatic rosary that's totally congested. You know, nothing hardly gets through, the colon is so layered with stuff. In autopsy, people with regional enteritis, their colon is just layers and layers of something. They don't even know what it is. They don't absorb foods, so they become thin and anemic and any disease can wipe them out and so I'm sure I've had a colon in this state for quite some time, plus tearing in the rectal area. It's a classic disease and it just showed in my eyes. But, back to the eye. It's so much different now."

Lavon: "How are you now?"

Gloria: "Well, there's not that heavy, heavy orange, showing on the iris of the eye, there's still some. Sometimes along the way I keep thinking, I've been on this a year, but I know people who have been on it four years and still have a long way to go, so I don't feel too badly. I have eaten a lot of the wrong combinations and a heavy protein diet because I always thought pork — lots of pork — and it's interesting, I've looked in the eyes of the people I've known from the South ever since, and I can't believe it. Their eyes are all orange. I know these people that raise pigs down in North Carolina, and I looked in their eyes and I couldn't believe it. Of course, I don't know if it were that exactly. But this one person's eyes I looked into, he's a good, good person. He's got all kinds of spots and orange. I notice eyes now."

Lavon: "You're thankful?"

Gloria: "Well, it's hard to say where I would be now. I know there was no way out. It was like a bottomless pit. This was the light I needed. I'm thankful, and if she listens to this tape she knows I'm thankful to her for what she's taught me."

**Loretta Adams**

This happened to my son who is seven years old, he had trouble all of his life since he was a baby with his bowels, and when he was about six he could only go to the bathroom about once a week and only with a suppository. I took him to the doctor, and he put him on mineral oil for two months. I had to give him two ounces every night of mineral oil, and it helped the condition. He got so he was going to the bathroom more, but I didn't notice until later, all but his vitamins and everything that he got from his food was going out with the mineral oil, and he got real bad strep infections that lasted for another couple of months after that. He would get them, we would get it almost cured, and then it would start back again. After he was off the mineral oil, he was all right for a month or two and then it started again, worse than before. He couldn't go to the bathroom so we took him back and the doctor took x-rays. In the lower bowel, in the sigmoid colon, there was a place that narrowed down to almost pencil thin, and the doctor suggested that we bring him back, schedule him for a biopsy, and then to do surgery to remove that section of his colon. He said he didn't know what caused it, that it might be a part of his colon where the nerves weren't working.

We had him all scheduled for February for the operation, and about three weeks before the operation, I talked to a friend of mine who had been to some classes and she told me about the herbs, the herbal laxative, and the enemas and the things to do for it, and I thought well, I'll try anything because he had had several surgeries already in his life, and I didn't want him to have to have another one. So, we gave him the enemas and the laxative. I gave him an herb laxative every day for a week and nothing happened. All of a sudden one Sunday as we were getting him ready for Sunday School, he got sick. He had diarrhea, he started throwing up, and everything. I thought he just had the flu, but evidently this laxative finally started working on him, and I kept giving him a laxative and the enemas and then he had some bleeding, too. Red, fresh blood, so evidently the enemas had caused the bleeding, so I gave him golden seal enema, and we worked on him like this for about three wekks. He was scheduled to go to the doctor for this biopsy, but they took x-rays first. I went over to the doctor's office. I had to go in first to talk over the x-rays with him before he went to the hospital, and he said, "I don't know what happened, but those areas are completely gone that he had on his colon." So he said, "I don't think there's any need to do the surgery or the biopsy, either one now."

**Doctor Richard Storres**

My name is Richard Storres. I'm 45 years old. Approximately five years ago I started to notice symptoms which I thought were of orthopedic origin, loss of function in my legs, which I later found out to be neurological and was finally diagnosed as multiple sclerosis. About three years ago I learned that I had diabetes and was started on oranaz, an oral insulin, which later was changed to regular insulin shots, and approximately two months ago I was taking as much as 300 units of U-80 insulin — two shots in the morning, two shots in the evening to keep my diabetes under control and even then my tapes would often run +3 and +4.

In the meantime, my loss of function progressed to the point where I had to use a cane after about a year-and-a-half and then a year after that I had to go to a walker. Just two months ago I was looking at the inevitability of ending up in a wheelchair permanently, and probably spending the rest of my life confined to a wheelchair.

Through four different people I learned about LaDean Griffin and her mild food diet, and obtained the book entitled, *Is Any Sick Among You*, which I read several times and decided to launch myself on her program religiously.

After just sixty days, I have lost 32 pounds from 190 to 158, and I am able to get around much better than I did before for a long time; I'm able to walk briefly for short periods at certain times without assistance whatsoever, but the most important thing to me is I have been able to for over a month eliminate the need for insulin altogether. My tapes are completely negative and have been for over one month. As far as my expectations for the future, I expect and intend to follow this diet on a permanent basis and to eventually be able to walk normally once again and to regain my full health.

Lavon: "Isn't that a thrill to know that that's possible?"

Richard: "It certainly is. I never thought it was, but I've seen dramatic results in just a short period of time. I can imagine what would happen over a period of a year or two.

Lavon: "Oh, yes, I really believe in it. I think that I was led. I've prayed about it a lot. I really believe that I was led to this source and to become acquainted with LaDean Griffin."

Lavon: "Well, that's a thrill. Thanks very much, Richard."

Richard: "Thank you."

**Shirley Story**

I have a lot of little things to tell about, but the one main thing is kidney stones. I had suffered from kidney stones for a number of years. I've been hospitalized twice with them, so I knew what they were, and I kind of felt like the Lord sent this one on a Sunday just so I couldn't get hold of a doctor. But, I had gone down to church about three months ago, and in church I started getting this real severe pain, and I knew what it was, and by the time that church was over it was excrutiating. I could hardly stand it. I went home and my husband frantically got on the telephone and started trying to call the doctor and couldn't reach him anywhere. In the meantime, I went in the bedroom and prayed about it for some kind of a feeling or an answer on what I should do. I don't know why. I hadn't even thought about LaDean's class that I had just taken last fall, but it came to me, anyway — LaDean's class — and that's all that came to me.

Lavon: "That was an answer to prayer."

Shirley: "Yes. Then I went to LaDean's book, and started taking the juice of a lemon every time the pain hit, which was about fifteen minutes."

Lavon: "Straight juice, or water with it?"

Shirley: "I mixed a little water with it about a fourth of a glass."

Lavon: "Almost straight?"

Shirley: "Yes. It was hard to down, but I made it. And I could see that the pain, it was getting less. I took this all day long until the pain was completely gone."

Lavon: "Maybe seven or eight times, you took it? Like every hour?"

Shirley: "Oh, I took it more than that. I took it every fifteen minutes for a good two hours.

Lavon: "Is that right? Did it start to let up right away?"

Shirley: "No, not for a while, not for about two hours. And, in the meantime, my husband said, "You've got to go to the doctor. We're going to take you down. We've got to do something." And I said, "No, I want to try this because all he'll give is pain pills." So I stayed on this all day, and by the time night came I was not in too much pain. I could stand it.

Lavon: "There was a relief?"

Shirley: "Yes, and the next day it was gone. I didn't have any, so I went ahead and I took the formula that she recommended for kidneys and then I stayed on just parsley — a parsley tea for about three days. I didn't eat anything, and I haven't had any more attacks."

Lavon: "That's wonderful, how many months has this been?"

Shirley: "It must be about three months."

Lavon: "Do you think you've passed the stone?"

Shirley: "I may have passed it or it was dissolved. That's what it does, it dissolves. She says in her book, it will dissolve it like water on rock salt."

Lavon: "Is that right?"

Shirley: "Yes. In fact, I take a little lemon juice once in a while when I think about it just to ward this off. It really works."

"My dad has Parkinson's disease. He went to an acupuncture doctor and had treatments and that didn't help. I've tried to talk to him and tried to get him on this mild food diet and raw juice diet and all the B vitamins, but he just can't see it. So, finally he decided to go to the hospital and have some treatments. He was on ladulca, and it worked all right while he was in the hospital. We could see some difference, but after he got home in maybe two months time he started getting really deathly sick. He couldn't keep any food down. He was going downhill, losing weight, and the shaking was just worse. His whole body shook inside and everything, so finally I thought, "Well, I've just got to face facts with him, you know, and say, 'Daddy, this isn't going to cure you, this isn't getting to the cause of it and you are wasting time.'' So, this is what I did, and we finally have him on a real high B complex vitamin, and he's taking brewer's yeast, vitamin C, calcium and magnesium. He's stopped taking the ladulca pills. He stopped taking them gradually. He doesn't take those now at all. His shaking has improved. He doesn't shake inside anymore. He says it feels so good not to be shaking inside. His thinking and everything has improved. He even got to a point where he couldn't drive, and his eyes were kind of starey. Now, he's driving again. Well, that's a testimony for what food factors will do. Another thing, cayenne pepper. I've got a boy who is grown now, but when he was small, he had asthma so bad that we'd have to put him in the hospital — oxygen — to breathe, and this would come on in the winter or the fall when it was real wet. He seemed to be allergic or something to the dampness. So I had a friend that was apparently into the herbs, but I didn't realize it at the time so what she said went over my head, but she said, "Shirley, please try cayenne pepper." I said, "Well, I don't know what to do with cayenne pepper," so she fixed me some capsules, and I gave him those. I can't remember what the dosage was. But I started those in the late summer, and he didn't have another attack, and we gave him those all through his growing up years and he has never had another attack.

Now, he's in the Navy. He's out on a ship where it's really damp, and he doesn't have any problem.

Lavon: "Does he still take cayenne?"

Shirley: "He doesn't take any now. Well, that just really cured him of that. He had it so bad that his ribs hurt and he would shake and try to get air." Then after he took cayenne for maybe two or three weeks, he didn't have any more attacks, but I didn't let up. I kept giving them to him through the winter, and then as he grew up, each year in the fall I would start giving them to him."

Lavon: "It's really great. Do you have any more to tell me?"

Shirley: "No, I don't think so. Yes, I do. Yes, I do. With the vitamin C. About eight months ago I had a real severe earache and, of course, again I wasn't using my head, wasn't thinking. Well, I ran to the doctor, cause it was so bad I couldn't stand it. He made a culture and found out that it was an infection in my inner ear, and it was throwing my equilibrium out of balance, and I was working at a mill at the time. I could hardly stand the sounds of the machines running, so I got on the medicine that he gave me. It seemed to clear it up, and then again in maybe two weeks it came back. I was talking to my friend about it, and she said, "What about vitamin C?" So, I started taking vitamin C, a thousand milligrams every hour, and within about two days it was cleared up. So, I keep lots of vitamin C in the house."

Lavon: "We should have a lot of C for storage. Is this the end of your testimony?"

Shirley: "Yes."

Lavon: "Well, thank you. That was great."

**Arvilla Cox**

My testimony has to do with colitis. I had problems in my colon and I even thought it might also be ovary problems. I had pain in the lower stomach quite a while, I had this off and on. I didn't have any idea what it was, but at this particular time I was so sick that I decided, "I'm going to the doctor and find out what it is." I had already taken LaDean's class, but sometimes you decide you could treat it if you knew what to treat, but you're not sure what you're treating, so I went to the doctor and they started running tests and decided, it might be colitis, but they weren't sure. They'd taken x-rays, so they put me on something — I don't even know what it was, it was ghastly. I came home and took it but nothing seemed to help. In these type of cases, you know, you can keep on hurting and then, back in four or five days and see how you're doing and if that doesn't work, then we'll try something else. So, this is exactly what I did. After four or five days I was still in bed. The fact is that I was so sore that for someone to sit on the bed that I was lying on, hurt. That jarring hurt me, and so I waited the five days and went back and they said, they had decided to try something else, so they gave me more medications and said, "Try this." So, I went home and this went on for three weeks. I was almost totally in bed or in the bathtub because I felt better in the bath, with heat or a hot water bottle. This is the way I was for three wekks and so I got up one morning and the doctors had just about decided it was colitis, so, I decided, I had taken LaDean's class, so what in the world am I going through all this for. I couldn't possibly do worse than the doctors. I decided this is the way it was going to go, so I got my slippery elm out. I had all these things, and I put the slippery elm in a capsule and I started taking it. I took it the first day, I can't remember the amount. It seems like it was about four capsules, every four hours.

Lavon: "Did you take okra?"

Arvilla: "I didn't even think about okra. I had slippery elm. I did real well, in fact, just that night I was feeling a little bit better, but I have to tell you, maybe this is the problem with all of us, we don't exactly stick to things. My daughter made a big pad of fudge the next day. This is one of the things, you know, diet, has so much to do with everything that we do. But I could not resist the fudge, so the next day I had a few problems. But after the third day on slippery elm I was completely free of pain. I didn't have one bit more. I don't know what it does, but I'm a zealot when it comes to slippery elm. I'm not kidding. I'd shout from the top of the housetops because I still have problems once in a while. I can feel it coming on. I really wouldn't have believed that it was caused from stress as much as it is until I'll have things that bring on stress, and then I'll find it comes back as quick as I put myself under stress. When I do, I take slippery elm and by the next day I'm fine.

We have a lady in our area that was having a problem and she'd been to the doctors. Someone said, "Go talk to Arva. She's a slippery elm gal," so she came and talked to me. She asked me what I thought, and I said, "I can only tell you it works if it's colitis, because I know about that." I talked to her about two days later after she tried it and she said she was feeling a lot better, and I've talked to her since and she doesn't have a problem. She had lost weight."

Lavon: "See how these testimonies help people learn?"

Arvilla: "I think this salve they're talking about — this cancer removing salve is really great. Do you know, I can testify to the things that I've tried and I know. I think the whole thing is great, and I'm sure that it all works. But you know for a certainty when you've tried things and they work. I know my little boy has problems with boils. He loves to ride horseback and he has to ride bareback. He gets an infection, a boil and this salve they've got is great for drawing. If I catch it when it's starting, within three days it's come to a head and it's over. In fact, you should have him give his testimony. If anything goes wrong, he says, "Just ask my mom. That works." I've got a few problems, but it was real funny when I was taking LaDean's class and I first started, and she started teaching us these things. In these things I found out that cayenne pepper would work for ulcers, and I was having a lot of trouble with ulcers at the time. This is another stress-type thing you know, and so this one night I was to go to class, and I hurt so bad, and I tried everything that I knew, like soda, and all the bad things that I tried and nothing worked and finally I called Bret and said, "Bret, if you've got some capsules, bring them with you, or else I can't go to class." I said, "We'll give it a go. If I'm going to die, it will be down there with LaDean," because I was so afraid of it. You know if you've got an ulcer and you've talked to doctors you do not take spicy foods, and so I thought, "Okay, I'm going to put this to the supreme test because I'm taking two capsules of cayenne pepper," and I took it, and I went down and actually forgot to tell LaDean that I had taken it because I quit hurting, you know how you do when you're not hurting, you forget, and then it just quit. It just quit.

Lavon: "I hear a lot of good things about cayenne."

Arvilla: "I found that too. In fact, what are these kits we're making in Relief Society? Safety kits? First-aid kits, and we've been told — nurses are helping us get these little kits ready, and we've got all these neat things to put there in case of emergencies, and in mine there is slippery elm and cayenne pepper and vitamin E. I have all these little extra things, these are the things we must store."

Lavon: "Well, Arvilla, I think you're testimony's fantastic. It's going to help a lot of people. Thanks a lot."

**Beryl Furner**

At the time I was studying and taking classes from LaDean, I conversed with my sister about what I was learning. Because of my studies in iridology, I began to notice people's eyes and my sisters's eyes were particularly interesting to me because she had one or two dark spots in each iris. I hesitated to talk to her about them, not knowing how she would accept it. After a few weeks I observed that one spot became larger and very dark.

My sister had complained of back problems and pain which she suspected to be gall bladder. Her doctor, after x-rays and checkup, told her he could offer no help or determination as to what the problem was or what to do about it so to let it go for a time. The spot grew larger. It was located in the right iris beginning in the ascending colon area and spread up into the shoulder area. The pain was in the upper right side of her back, under the shoulder. After hearing in class about a man with a similar problem and what he had done for it, I suggested to my sister, go on a mild food diet, take psylium seed powder in water with laxative every night and also take a high enema each day for at least three days.

She continued with this treatment for two weeks. She noted the spot in her eye had faded. She told me she thought she had passed a bowel impaction. When I saw her after two weeks I detected no spot left in that area of her eye.

**Gloria Wood**

About my finger. I stuck my finger — my hand — in the sump pump, and it cut the fingers, the one real bad, and it was sewer water. It was open, so it got an infection in it. We cleaned it out and used this cancer formula, and it has done wonders. It drew all the infection out. We just kept putting it on it. It left a scar, but it closed it right off and helped it beautifully. In fact, as my husband says, "That would make a believer out of anybody." Just a few days and it was great, well about a week, it was a deep cut, and it really did wonders. I have two boys that have ulcers and I have ulcers. We used slippery elm and okra, the minute their stomachs even think they're going to hurt. Another one is golden seal and myrrh. That does wonders for trench mouth or cold sores or canker sores. My son had trench mouth so bad, he had it clear down the back of his throat and all out on his neck. He couldn't even eat, he had to suck through a straw, it was so bad. He rinsed his mouth with golden seal and myrrh and it just did wonders.

**Veva Whitear**

Lynn: "Veva, we'd like to hear your testimony. We hear that you've had a little girl and she's been much improved since you've had her on mild foods and licorice root."

Veva: "Our little girl has hypoglycemia. When I first suspected she had it, she would wake up in the morning when she was just a baby. Then I'd give her a bottle and her little hands would shake so bad that she could hardly hold it. She needed food. Her blood sugar level was low. When she got about three years old, in the morning when she woke up if we didn't have food immediately the first thing she would do would be shake. The next thing she would do is say, "My stomach hurts." The next thing she would do — I knew I had about ten seconds and she would feel like she was going to vomit and then she would pass out.

When we went to the doctors, they kind of pooh-poohed this hypoglycemia and they wondered why I thought this was the problem, and I told him because it ran in the family. Her father, some twenty years ago, had a problem and he had done the same thing in the morning when he'd been without food all night and the blood sugar level was low. He'd go out and milk and do chores before he'd eat any breakfast and if I didn't have some food for him in a hurry all at once he'd become terribly ill and he would pass out. His family thought he either had a tumor on his pancreas or it was hypoglycemia. It turned out to be hypoglycemia. But when I took my little girl to the doctor he said, "Well, what makes you think she has this problem?" I said, "Because she reacts just like her father reacted some years ago; he doesn't have the problem now, apparently he is balanced now. His sister has the problem also." They pooh-poohed it. They said, "Little children don't have this so much and it's kind of a fad thing these days."

One day in a dentist's office she got scared because she's afraid of the sight of blood. They had to pull a root of a tooth out that she had broken off. She became so upset she passed out. So, she doesn't have normal balance and cannot take stress. I took her immediately to the doctor, a pediatrician — and he said, "Oh, I don't think we need to worry about that. When I was a kid I used to pass out when I got scared, too."

At Christmastime she had had a bunch of sweets at a party at school — this was when she was in kindergarten. When she came home she was running around the table with her little friend who was visiting and she fell and hit the corner of the stereo right on her knee and it hurt her severely and she went into convulsions. I decided then, "Oh, boy, we've got real problems." Passing out and convulsions and I didn't know what this convulsion bit was. I called the doctor and he sent me to a neurologist. When I went to him they put her in the hospital, and ran tests, costing us a lot of money, and they couldn't come up with anything that was wrong with her. I kept telling him I knew she had hypoglycemia, and he said, "This convulsion thing — well, we don't know what that is. I have about ten children a month and they hit themselves and they go into a convulsion. It doesn't happen to older people, it happens to children." But he was concerned about this other — passing out and so forth. When he got through with the tests he said, "Well, it didn't show anything." He was worried about epilepsy, it didn't really show this by the tests he took, but he put her on dilantin. I was just so sure that he hadn't come up with what was wrong and I knew what was wrong. I went down to the health store and started talking to them. I got Adelle Davis' book and I started reading in there and it gave all of her symptoms, even to the part of convulsions when they have this problem, and so I threw away the dilantin and started her on some vitamins at that time. I hadn't met LaDean then. I found our neighborhood had classes that LaDean taught.

Lynn: "Did the vitamins help a lot, though?"

Veva: "Yes, they did help a lot. It wasn't the whole answer yet. She was still shakey every once in a while, and I had to be very careful with her, and when there were stress times when she was really frightened, I had to be careful. After I took the class from LaDean and found out about the licorice root, I just started her with licorice root. I wasn't convinced at this particular time, but I started her anyway."

Lynn: "Is she taking any other herbs?"

Veva: "No, she's taking vitamins. She takes potassium, magnesium, vitamin $B^6$, B-complex, etc."

Lynn: "Is she taking a lot of C?"

Veva: "No, I haven't given her C. I was giving her some help. I thought this wouldn't hurt, but, you know, she got high-strung and she got really nervous. I thought, "Something I'm doing is wrong." I kept thinking about kelp having to do with the thyroid, and I thought, "I wonder if she's getting too much and making her nervous." It is natural thyroid, and I took the kelp away from her and she was fine.

She did so well for a while, we kind of decided, well, maybe she's outgrowing this, so after a while, we ran out of vitamins and I didn't get them. She did really well for about six months and we had no more problem. We thought maybe she was over this problem. Last March she and an older sister decided they wanted to get me and my husband an anniversary present, and they didn't have the money, being little children, but they wanted to give us something nice. They decided they would go without their lunch at school to save their lunch money for the gift. Well, the first part of April she had been going without meals which she shouldn't do with low blood sugar. One morning she got up, she hadn't had a problem now for three years. She was practicing piano and all at once she jumped up and started running to the kitchen. She felt a pain in her stomach and she knew, she just had a few minutes to get something in her stomach or she would pass out. As soon as she got to the end of the couch she fell on the floor. She tried to get up, but she couldn't. All at once it dawned on me, "Here we are." I ran to the kitchen to get something fast, but it was too late. She said, "I'm going to throw up." Then she passed out and went into a convulsion. This was after she had been going without meals for nearly a month."

Lynn: "She learned a lesson there, didn't she?"

Veva: "Really, and so did we. We got her back on herbs. She is really on a program now, and doing fine."

**Jan Perry**

We've been on mild food about seven months now. We really feel like it's helped us. I for myself, feel like this method has saved my life. I was told I had leukemia and I feel now that it is on the mend.

Lynn: "Have you been on it strictly, the mild foods?"

Jan: "Pretty strict. We've eaten a few things now and then, as most people probably do, pretty strict on mild foods. I've used psyllium, the CS formula, liquid dulse, licorice and lobelia."

Lynn: "Have you noticed a big change in your eyes?"

Jan: "It's really hard for me to compare my eyes with the slide. I'm anxious to have our eye slides taken again and compare. I feel like there is a change in my eyes. I feel so much better than I did before I started."

Lynn: "Did you have any specific problems, other than

leukemia?"

Jan: "I've had female problems and when I look at my eyes, I wonder what I don't have."

Lynn: "How's your family taking to this?"

Jan: "My family really has accepted this well."

Lynn: "Have they done pretty well, too?"

Jan: "Yes. One problem we have is they're tired quite a bit, but I guess that's due to eliminating. At least, I feel like they are. I'm really glad that my whole family has accepted this.

Last summer I went to the doctor and my blood was real low, and for a period of about three months I kept going back. He'd check it every week, every two or three weeks, and he kept giving me iron shots and iron pills and all kinds of things, and he was very alarmed that my blood didn't come back up. At this time I was told that I had a leukemia condition, and that's when I started on the mild food, and I have felt so much better that I really have a testimony that it's right."

**Elaine Thompson**

I have a boy who is retarded and he's also epileptic. When he was two days old, he started having seizures. I didn't get him out of the hospital until he was two weeks old, and he was put on phenobarbital. We had him on that for awhile and phased him off and he was fine. He didn't have another seizure until he was eleven years old, and then he started having them again. I took him to the Oregon Medical Center. I've had him to several neurologists. I've had him to one in Ogden. I've had him to the University of Utah Medical Center, and no matter what they gave him, it didn't control it. He was so drugged at one time that he couldn't walk a straight line down the hall at school. It just made me sick to watch him because he was — well, you know, it's bad enough when he's retarded, but having this on top of it. It really bothered me and — so, when I heard about this class of LaDean's, I thought, "Well, gee, that might be someway to help him."

Lavon: "Do I understand you now that you've had him on drugs up to this point?"

Elaine: "Up to the point, yes, where I came to LaDean's class. In fact, he was taking dilantin and then he was taking something stronger than that, and I can't remember what it was. He was also taking something stronger than the phenobarbital. It was about the strongest things they could give him. He was taking about two or three of them.

Lavon: "Did he act drugged up when he took those?"

Elaine: "Oh, yes, he was really, really drugged up. Besides that, they have a lot of trouble with their gums with the dilantin, unless they keep their mouths real clean. I've always had trouble getting him to brush his teeth. But anyway, I took the class and so I decided I would take him off the drugs, the dilantin and the other drug. I replaced it, four to one — four nervine herb tablets and one of the other kind and then I gave him B complex and calcium — four of each to one of the white pills that he had. As I upped the nervine and B complex, I lowered the other, so I didn't do it all at once. It wasn't just a straight cutoff. I combined them at first and then I gradually took him off all of the drugs. He's been off them for more than a year-and-a-half.

Lavon: "He isn't a bit worse?"

Elaine: "No, he's really better. When I first got him off of it, when he'd have a spell, he'd really be sick and I decided maybe it was withdrawal from the other drug, because he'd be sick for maybe twelve hours after. He was sick to his stomach and would throw up and now he's gotten to the point though he still has some seizures, he has them mostly at night and he doesn't get sick like he used to. He still has to sleep them off, but he's so much more alert and he's so much more easy to work with because he isn't drugged up. This fellow I know is drugged up all the time. He just looks like it. You could see the look on his face. Well, that's the way my boy was."

Lavon: "Then you really are thankful for this way of life, aren't you?"

Elaine: "Oh, yes, I wouldn't go back to the other for anything. I think if I can get him on the right diet. He's eighteen now and he doesn't like to be on a mild food diet. I think if I could get him on that kind of food that it would lessen it a lot.

I also have one other thing to tell you. I have a daughter that gets earaches quite often. About a year ago, she had a real bad ear infection. In fact, she couldn't stand up. It seemed like an inner ear infection, so I just kept her on high vitamin C, and I rubbed her glands. She also had swollen glands and I rubbed them with an iodine salve and kept cold packs on her ears, and she was over it in about two or three days, which was probably a shorter time than it would have been on anti-biotics to get over it because an inner ear infection can go on for a week or two.

Vitamin C is a fantastic thing. This was a thousand units every hour, and whenever the kids are sick, that's what I say, "O.K., a thousand units every hour." Vitamin C and juice. I forgot to say I've got Gary on Vitamin C three thousand milligrams a day, he takes B Complex and $B^6$, magnesium and just about everything. Well, just about anything I hear of that's helped an epileptic person I start giving it to him and I think they all kind of help him. He used to be quite sickly, but now he's hardly ever sick."

**Hawley Haws**

I'm just beginning to learn a little bit about herbs but what I have learned so far I'm really excited about, because I'm able to help myself and my children and my husband without running to the doctor the first little thing that comes along. First of all, when my last little baby was just tiny, I brought her home from the hospital, she had a rash on her bottom right from the start and I put paper diapers on her, so it wasn't a matter of diaper rash. It was some other type of rash she picked up at the hospital, and I tried everything. All the normal creams that were on the market.

Finally, I went back to the doctor because it was just a blistery little rash and very sore and he gave me a prescription which was a $7 or $9 tube with about a quarter of an ounce of cream and that would begin to almost burn the skin till it was just kind of cauterizing it, but then the rash would come back immediately after. So that wasn't doing a thing. And then I read in one of the books that I have — I don't remember what it was — that golden seal was a real good thing for skin rash and other skin conditions. I used it just on the clean bottom all by itself, instead of baby powder, and overnight the little bottom began to dry up and the blisters cleared up within the next couple days. Now whenever I have a problem of the rash coming back at all, I just put a little of the golden seal on and then leave the little bottom open to the dry air during the day as much as possible, and it really does the trick. It heals it so quickly, I was amazed.

We've also used golden seal on canker sores and other skin lesions and then I discovered, thanks to my Aunt LaDean's book, that catnip tea is real good for colic, and I've used it with the last couple of babies and had almost no problem with colic. The catnip tea tends to be just a little bit bitter, so I put a little bit of honey with it and a little bit of alfalfa mint tea or another mild tea with it. I've also used comfrey tea as a cleansing tea for my smaller children to help get rid of colds or flu quickly because the other teas tend to be a little bit strong for their little systems.

Now the real thing that was a testimony to me about herbs and vitamins and cleansing diet, was when I got hepatitis, a rare type called Primary Cholangiolitic hepatitis. My whole skin would itch and I turned yellow, of course, but the main problem with hepatitis is that it can cause permanent damage to the liver and kidneys if it's not taken care of properly, and I knew I had to have complete bed rest. Since I had four small children, my mother came and took care of the two smaller children while I just stayed completely in bed. Then my sister took the two older children with her. I had heard that with this disease you must have complete hospital rest for six weeks to two months. I expected that I would have a serious problem getting myself back on my feet. In fact, with this kind of hepatitis, I was told that there would be approximately two years to recuperate, where I would have my complete strength back, and I knew that I couldn't afford that with all my family's needs.

With the help of my mother and Aunt LaDean's book which we consulted because she knew more about herbs and about nutrition for ailments than my mother did, I began an herb program. My mother knew enough to put me immediately on a complete fast first and then gradually I would have vitamin C at high potency and with help from my Aunt LaDean's book, I took lemon juice in water with no sweetening in it, about every hour during the day, along with the high potency vitamin C about every hour, I had about 1000 to 2000 milligrams every hour. I had vitamin A and vitamin E and I took vitamin B complex because I was taking golden seal and they need to work together. I also was taking the CS formula because they are excellent cleansing herbs, and this is what I needed to do, get that all out of my system before it could do any damage to my organs. I also took another formula, I don't remember the name of, especially for cleansing the liver and kidney which were the two that were affected the most by the hepatitis.

It was really amazing. I knew that I would be over the hepatitis enough to get out of bed soon. The yellow was completely gone and out of the whites of my eyes as well. Within about two-and-a-half weeks that was completely gone, and here I had expected it would go on for six weeks to two months, and then within another week I was completely back with my energy back and my strength and I gradually brought fruits and vegetables into my diet and now have my full strength. I am expecting our fifth child and I feel better with this pregnancy than I remember ever feeling before. I've had no morning sickness, and I really think that the cleansing has helped this pregnancy to be better and to be easier. I'm really happy that I have Aunt LaDean, and that her book could help me and that I had the help of my sweet mother and sister.

**Bessie Upwall**

For the last four-and-a-half to five years, I have had a real bad colitis problem. In June of 1974, I was advised by two doctors to quit my job because of pressures. I wasn't getting enough rest and was working too hard. They advised me that I would never be any better if I didn't give up my job. I quit my job for a year and a half and it was then I met Lavon Thomas. She introduced me to herbal living. I read LaDean's book about herbs and decided that that was the way I wanted to go.

I started taking herbs four or five months ago. I took saffron, slippery elm, black walnut, psyllium, the CS formula, capsicum and I've also been taking a combination of hops, valerian root and scullcap. I think that has helped me more than any,if I were to say one herb combination was the one that really helped me the most. It has helped my colon to remain calm along with everything else. Then I went on a strict diet, left meat alone, ate vegetables, fruits and so on. I not only got my colon better but I lost 13 pounds. I'm still trying to eat right. If I cheat and eat some things I'm not supposed to then I suffer a little bit but nothing like before.

I'm not holding down a job again, which the doctors said I probably never would be able to do, having lots of energy, being able to do the things I need to do and feeling better than I've felt in five years.

Lavon: "Isn't that great. You attribute it all to the herbs?"

Bessie: "You bet I do because I haven't felt this good in that long and I know that nothing else has helped me before.

Lavon: "Do you feel that in another month or two your colon will probably be well?"

Bessie: "Yes and I might add too that I was going every three or four months to Wyoming to a doctor and he was doing adhesion work on me. When I was there in December, I had only been on the herbs two or three months and he said that it was better than it had been in years. I hope to be able to lengthen out my visits now and possibly not have to go back again."

Lavon: "So you're a real herb fan huh?"

Bessie: "You bet and I thoroughly believe in capsicum because boy that really is a healer and sure gives you lots of energy.I take it faithfully."

Lavon: "Do you?"

Bessie: "You bet. Oh, I might add another thing about black cohosh; also in 1974 — in June — I had a breast that was weeping and the doctor had to cut me way back on my estrogen and said I may have to go off of it completely. I wasn't able to go off of it completely because of other symptoms I had then. When I started on black cohosh, I was able to cut my estrogen way back and take the black cohosh and now my breast has completely cleared up."

Lavon: "Is that right, and how long did you have that?"

Bessie: "I had it for two years."

Lavon: "So you think black cohosh did it?"

Bessie: "Oh I know it did because its helped to balance my estrogen and let me cut way back on the other estrogen product that I was taking and I've felt great. I haven't had a lot of hot flashes or depression or anything like I had before."

Lavon: "And you've really been through a lot of things too."

Bessie: "I sure have."

Lavon: "A lot of stress?"

Bessie: "Yes, a lot of stress and strain, I've been going through a divorce."

Lavon: "Your testimony is going to mean a lot to people that read this book."

Bessie: "I hope so because I know that without herbs, I would be where I was several months ago, only worse, because of all the stress I've been under. I would really have been down the tube if it hadn't been for those herbs."

Lavon: "One thing that I can appreciate about you Bessie is that you take them regularly and the way you should and drink plenty of water. If people would just stick with the program, they would see results."

Bessie: "That's true. I found that if I stay with it really faithfully, I just feel great and I get along good. Like I said, if I cheat on it, I eat sweets or if I eat meat very much — once in a while if I eat a little chicken or something like that it doesn't bother me, but I try not to do that."

Lavon: "Eat a beef steak or anything?"

Bessie: "I haven't had a steak in months and months. If I eat anything that has very much meat in it, I really suffer, that or heavy sweets. So I know that those two things I really have to leave alone or eat very, very sparingly. If I just stick to fruits and vegetables, I just get along great."

Lavon: "Ok, thanks a lot Bessie."

**Linda Beal**

When I was young I didn't pay much attention to the Word of Wisdom except for no smoking, no drinking. Always when I read the Doctrine and Covenants section 89, I felt that I was doing right, meat could be eaten anytime, herbs and that advice were for people in Moses' time.

As I went through my 24 years of life, ate all things that I wanted, when I wanted, I had all childhood diseases. I was always sick, with sore throats, ear aches, constipation. I had terrible cramping when my periods would start then less period and more cramping, more often — three times a month. I used a bottle syringe for a full year because I didn't have any other way of relief of constipation.

My cousin told me about mild food diet and how it could heal your body. I believed everything she told me that day. I started immediately, left my second glass of cocoa and four pieces of toast and lunch of Dr. Pepper and candy bars alone. I thought, "Really

I'll eat better now than I ever have." I had had a terrible pain on my left side and went to the doctor. I had x-rays, wasted a lot of time and money and he said, "Can't see anything wrong with you." Something had to be wrong, you don't have three periods a month and constipation that bad for no reason, no wonder I felt sick.

I ate mild foods and took herbs, psyllium, lower bowel herbs, CS, female corrective herbs, nervine herbs. Miraculous things began to happen immediately. Bowels were already working better. I don't use anything like a bottle syringe anymore. I have good periods and they are regular. I felt like I was alive with lots of energy. I could read and not fall asleep.

I lived for four months on mild foods. It took me four months of living like this, then my body went through a healing crisis. You really feel sick with it. It lasted only one day and I felt like a million the next day. I try and live on a mild food diet and take herbs occasionally. I feel really good now. I know its the truth.

**Rich Wheeler**

By the time I was 24 years old in 1962, refined carbohydrates, principally sugar, had severely damaged my health. I had developed hypoglycemia. Unfortunately, it was a full thirteen years, before the correct diagnosis was finally made. Those thirteen years were filled with crushing medical bills, dozens of "specialists" and myriads of negative test results.

Prior to that time I was a regular guy; happy, fun loving, always quick to laugh. I seldom worred about anything. You can imagine how disconcerting it was for me to find myself constantly worrisome, and fearful. Equally disquieting were the ever present headaches, mental confusion, abdominal cramps and inner trembling. I was sure the local doctor could help me find the problem. Many x-rays and several dollars later, the physician indicated that the promised ulcer was not there. Nervous stomach was pronounced the culprit. However, the symptoms didn't leave as readily as my money had. The answer was obvious, try the specialists in the big city.

I went the full route. There were chiropractic adjustments, eyeglass fittings, dental spacing plates and tests and tests and tests at the clinic. The medics pronounced me clean, gave me a bottle of tranquilizers and wished me well. Think of it; my very first bottle of tranquilizers!

The summer of ‘63 found me newly married and living in Tucson, Arizona. The tranquilizers were ineffective and I kept driving my tired body as best I could wondering, “why me”? In August ‘64 an expert diagnostician, of which there were plenty in Tucson, prescribed Bellergral space tabs to alleviate my symptoms. It seemed to help but success was short-lived.

Acute pancreatitis landed me in the hospital in September. I was elated to have a real disease. It was good to know at last the cause of those terrible pains in the abdomen. Once the crisis was over I was dismissed without even so much as a word of counsel. The stage for the next crisis was thus set.

Early in 1965, two different cardiologists heard suspicious noises in my neck. An arteriogram was ordered and three days later the customary negative results were given to me. After contracting hay fever in ‘66 the summer of ‘67 introduced me to acute prostate infection which was to plague me for years. The Bellergral tranquilizer carried me nicely through the spring of ‘69 but I hated my dependence on it. Instinctively, I had cut down gradually and eventually decided to stop altogether.

During the summer long working hours, extreme mental stress and poor eating habits reduced my weight drastically. In October it happened! Uncontrollable body tremors and anxieties that could only be described as pure horror. Once again I reached out for the arm of flesh for help. Luckily for me my physician was acquainted with the “best” psychiatrist in Salt Lake City.

Along with psychotheraphy, the doctor tried me on several drugs to take care of the anxiety-depressive reaction I had experienced. In spite of the fact that no insomnia was present a powerful sleep inducing drug called Parest was administered. I slipped rapidly into depression. For over a month I was apathetic, lethargic and constantly obsessed by suicidal thoughts. Sinequan, a new drug, was given because it had “uppers” and ”downers” both. But the mainstay was to be the wonder drug, Valium, relatively new but unquestionably safe. Valium and I were to see many years come and go together.

In 1970 I began to lose a great amount of hair. At the same time I came to enjoy the toxic effect of the Parest even though the doctor had cautioned me about it. My body weight moved up to a peak of 200 pounds then slowly descended to my normal of 175. By the summer of '72 I was in terrible physical condition weighing only 168 pounds (I am 6'6"). To correct this I began to work out with weights and take a high protein anti-stress formula a la Adelle Davis. The results were impressive, I gained 20 pounds of good solid muscle and rapidly increased my muscular strength. The only problem was the growing intensity of those same abdominal cramps and raw eczema on the hands and feet. Before I could take evasive action, the pancreas inflammed again and I was hospitalized. I was sick with worry. After dismissal from the hospital I found that concentrated use of the B complex vitamin cleared up the hands and feet in about two weeks.

There was no improvement in my general health but with the help of Valium I was at least able to hold onto my job as a landscape architect. I was never far from the "wonder drug" and would never leave the house without patting my pocket to see that my friend was there. I was just plain hooked! I wanted off but the doctor discouraged me and also I lacked the personal courage and faith to do it.

Eventually, however, I was forced to resign my government post after twelve years of service. The quality of my work was high but the quantity was too low, the boss said. The truth was that I just didn't care about my work anymore.

In June '74 the prostate flared up again. We treated it with sulfa drugs. Then, to "take care" of my hay fever, I took a Kenalog shot. I felt fairly good through the summer. Pancreatitis struck again early in September and down I went once more. A week later the prostatitis manifest itself again. Tests indicated that I was anemic because of a shortage of hemoglobin. I experienced general apathy, fatigue and depression throughout that fall. The prostate infection submerged and then re-surfaced in April. Antibiotic drugs were now totally ineffective and my strength waned. I realized by mid June, '75 that I was seriously ill.

As the impending crisis loomed before me I knelt in prayer before my Heavenly Father to seek His counsel. Much to my amazement and chagrin, the Lord specifically instructed me to "get off medication". I was shocked! Didn't He realize how sick my body was? Why did He wish an additional problem on me at a time like this? I knew that I was addicted to Valium and that getting off would mean additional stress to my failing body. To say that I was frightened would be a gross understatement of fact. After wrestling with my faith for a couple of days and "delaying" my usual dosage just long enough to experience discomfort, I felt my cause was hopeless. The Father, always patient, spoke to me again through one of His servants. I was told that as I gained control of my own body, the desires of my heart would be granted. I knew then that He stood ready to help me accomplish that which He had commanded. I promised Him that I would stop taking the drug.

Hell yawned wide and I was swallowed up. Hallucinations, phobias, muscle twitching, pain and no sleep were the order of the day as my body cried out for its accustomed poison. Fortunately I was led by the Spirit to a doctor who ministered to his patients with natural extracts, vitamins, minerals, herbs, and mild foods. He was very helpful in this struggle to preserve my life. In addition to drug withdrawal and the underlying infection throughout my body, I had developed hemorrhoids, a protein deficiency, several mineral deficiencies, hypothyroidism, sluggish adrenal glands, digestive insufficiency, anemia, back pains, numerous allergies and, of course, I still had the ever present low blood sugar. Together with my faithful wife, I prayed day and night for deliverance.

Even as hell yawned, the windows of heaven opened above us. Blessings poured down upon our heads beyond our ability to contain them. For two weeks we were taught by the Lord. Many of the spiritual gifts mentioned in the scriptures became ours. Where doubt had existed, faith now stood. We knew that He lived. Being driven to our knees, as it were, the Father took our outstretched hand and led us daily.

It took another four weeks for the drug to be cleansed from my body to the point where I was reasonably comfortable some of the time. I experienced crushing headaches throughout the balance of the summer of '75 and the prostate problem gave me little relief. I grew impatient with the Lord and pled for Him to heal me immediately. I knew He had the power to do so; I just couldn't understand why He didn't.

The mild food diet was tough to stay on. Old habits crept back and, while I never touched refined sugar, I did occasionally take some meat and practically lived on whole wheat bread at times. However, a bounteous fruit crop supplied me with lots of fresh produce in season. I enjoyed, in turn, the cherries, apricots, pears, peaches, prunes and apples, consuming large amounts. As fall neared I was strong enough to engage in a few physical activities. Then the adrenals sagged even lower and back to bed I went. I found that the adrenals were particularly weak in the late afternoon, usually creating a crisis. Then about 1:00 to 3:00 a.m., I would awaken as they hit their low point of hormone production. At such times, there was difficulties in breathing and my chest muscles seemed partially paralyzed. Life wasn't rewarding at all. All this was attended by much mental confusion, depression and I often found myself crying at the least provocation. I wondered if I would ever really get well; I guess my faith was near an all time low. I had little or no patience with my children and wasn't particularly kind to my wife.

Since I hadn't worked at a regular job in two years, our financial situation was precarious at best. Miraculously, the Lord opened ways for us to be provided for and we never have been deprived. Our housing was less than adequate so as winter approached we began to look for a new home where it would be possible to juice vegetables, bake bread, sit in the sun and do other things necessary to getting well.

In December we moved to Southern Utah. After getting settled and after having recovered from the extreme stress of moving, I became really serious about my mild food diet. Right along I had been faithful in taking my adrenal cortial extract, licorice root, golden seal, and various cleansing and rejuvenating herbs as well as large amounts of vitamin C, B complex, etc. At Christmas time I started pushing lettuce, celery and carrot juice. I was very kind to my pancreas and ate mostly fresh fruits, vegetables and raw nuts. After about two to three weeks on this regimen I began to notice a strange thing happen. My mental powers were returning! My intellectual appetite became enormous and I began to read more and more with increasing retention of what I read. Even more exciting, however, I sensed my spirituality growing. I now realize, in retrospect, that the Spirit of the Lord certainly cannot operate to any great degree in an unclean tabernacle. Even though my body was getting stronger, I recognized that improvement in my physical health would be a pace or two behind the mental and spiritual. That was fine for me for now, I realized that I would eventually be completely healed and enjoy a degree of health hitherto unknown to me. For certainly, God cannot lie.

I know that all of my problems as outlined above came upon me because I violated the Laws of God. First, I ravaged my body for several years with excessive refined carbohydrates like sugar and white flour. Next I didn't meet the results of that mistake with faith but literally "worried myself sicker". Thirdly, and this also pertains to a lack of faith, I placed myself in the hands of uninspired doctors who almost destroyed my bodily organs with drugs through ignorance. I never plan to repeat these costly errors but, at the same time, I am joyous over the spiritual experience I gained.

Looking back at those dark days, I realize that the Lord spared my life by directing me to get off of drugs. I am grateful He loved me that much. I am thankful that He had enough confidence in my wife and to guide us to many truths. I have great joy when I reflect upon the experiences we had together, many of which are too sacred to relate herein. Out of these experiences I came to know and know now that He lives! He lives! I know that He lives and that Jesus is His Christ, that these revelations are open and available to all who seek Him. I know that if we will develop the faith and the courage to try His ways that we can all have these gifts.

The Lord has not placed us here on the earth to wander in ignorance. His Law of Health is available to us through the scriptures, through inspiration and through His chosen servants. If we will but heed that which we have been given and not violate the laws of God, we will have health in the navel and marrow in the bone. This I know. This I can never deny.

I am grateful for this testimony that I have, and pray that I might never do anything that would violate it and that I might endure to the end with my family and be exalted in the Celestial Kingdom of God. This I say humbly, bearing witness to all that I have said, in the name of Jesus Christ, Amen.

**Sandy Eggerson**

I am giving this testimonial to show how people can improve their health without having to go through all the operations and drugs. I've been ill for most of my life and not able to find out what was wrong; going from doctor to doctor. I have had several accidents and never any extreme illnesses just not well.

I went to LaDean's classes before her book was even printed or published, the first series, that she ever gave. It was on herbs and then I started on herbs and juices and various things. I started cleansing at different periods of time. Then I went to the iridology classes and she discussed how you could find different things in the eye. I decided that I would try some of the things that were useful for the different places that I saw in my own eye. One of the. main things was parasites and there was indication of possible cancer. The first thing I did was use the CS formula. I used oil of wintergreen, taking four drops each day for ten days and abstaining for twenty days and then taking it again for ten days. During this time I ate a fairly normal diet of vegetables and fruits and some bread, some cereals. During this time I passed at least three different growths, cancerous type growths that we could tell for sure. Then numerous, I couldn't say how many, probably at least a hundred round worms and five or six tape worms.

Lavon: "Sandy, what we want to stress now is that you left meat alone didn't you?"

Sandy: "Yes."

Lavon: "And how about sugars and starches?"

"Sandy: "Oh we've been off sugar for several years and no meat."

Lavon: "Did you go on the cleansing herbs?"

Sandy: "Yes and I used cayenne, the CAC formula, Lobelia and black walnut. The herbal pumpkin didn't do as well for me. I just was too full of toxins and poisons and so I had to use garlic enemas to help clean my body to help get the poisons out so that I wouldn't cleanse too fast. I don't know how many masses of worms etc. There was one time that I had a blockage because a mass had moved in the colon but we were able to get that out too. I wasn't even able to walk from the house to the garden. I wasn't able to do housework or even my dishes, it was just awful.

Lavon: "Strings of mucus came forth?"

Sandy: "Oh yes but the worms were fairly whole. I had the ability someway to cast them off and the growths too, the cancerous growths, came out most of them in one mass. Some of them had to break and rupture before I could pass them because they were too large. Your body can take care of just about anything if you can work with it and do the right things for it. I am very convinced that you do not have to go through all the drugs and operations unless you have been so seriously injured in a car accident, like a leg was amputated or something like this where the body cannot handle it where there is just too much destruction done. You can't expect that kind of a miracle but if the body is able to function, and my liver wasn't functioning. My kidneys were doing fairly well, the pancreas still is not functioning properly. I had sever hypoglycemia."

Lavon: "You used the licorice?"

Sandy: "Yes, I used licorice root and don't even need it anymore. I have gotten to the point where my body is working on its own and the licorice root just isn't needed now."

Lavon: "You couldn't go strictly fruit either, you would cleanse too fast."

Sandy: "No, I couldn't because I nearly died. I was drowning in my own waste."

Lavon: "So the mild foods is better to help to slow that down so you won't cleanse too fast?"

Sandy:"Yes.We don't use any type of dairy products now and we don't used meat and no sugar or starches, things like this. Everyone is a lot better."

I'm very thankful for the whole thing, for LaDean and for her books and all that she has researched.

*A profound study of the law of cause and effect is the foundation of wisdom.*

*Gibby*

# CHAPTER FOUR

## IT WORKS

*Know ye not that your body is the temple of the Holy Ghost which is in you, which ye have of God, and ye are not your own?*

*I Cor. 3:17*

*If a man defile the temple of God, him shall God destroy, for the temple of God is holy, which temple ye are.*

*I Cor. 3:17*

Through small beginnings, such as giving proof of the use of mild food and herbs by what takes place in the eyes, could bring our society to a point where organic food would be available everywhere, followed by herbs and natural vitamins. The knowledge of correct food combinations and sound principles regarding diet would then be taught in the schools. It could also be that sickness will be looked upon with contempt rather than with sympathy as it was in ancient Israel when sickness was considered a sin. We may then be asked, when we are sick, "What have you been eating?"

Following a lecture I gave in Los Angeles recently a man came up to me after the class and said, 'I've heard you say this so many times in your classes I thought you would like to have this." He handed me a badge with two words on it, "It Works." After giving it some thought, I realized that I was saying, "It Works", repeatedly simple because it **does** work. My own personal experience in addition to the experiences of hundreds of people over the many years I have studied the human body and natural methods have not always witnessed to me **how** it works; but with great conviction I have witnessed that it **does** work. Recently, by way of fumbling around for an explanation of how it works, I have heard people say — and even some have written — that it works merely on the principle of faith or belief in the herbs and mild food, or faith in whoever is teaching these principles. We have not looked long enough nor intelligently enough at how the body feeds and eliminates if we are willing to brush aside so great a truth with an "It's all in your mind" attitude. When the same herbs have worked for the same ailments for the many generations since written records have existed, and because they work equally as well on babies as they do on mature people, it logically follows that a change must occur, a final result that is caused by the correct answer to a given health problem.

Mathematically, sociologically and scientifically there are laws which dictate a given effect for a given cause. Why would the universe exist in such beautiful order if the human body were just a hodge-podge of chance? Some who live on the premise that the mind can do anything by just thinking positively, still die; and it would be with difficulty that they could grow back a leg that had been cut off, no matter how much they thought about it. The problem is that everyone becomes "too soon old and too late smart." By the time a family is raised the knowledge of how to raise children is grasped. Relatively late in life one begins to bring correct values into better focus. However, the problem of computed habit patterns in the brain, which started out like a fine thread, have been woven into a huge rope too difficult to break. Sometimes the only thing that makes it possible to sever this tie to bad habit is fear or pain, so that when the time seems ripe and the body is racked with pain as well as threatened with death, the struggle to loosen the body from the clutches of broken law begins. This is exactly what it amounts to — broken law. Perhaps not all the laws were broken that have brought the body to such a place of degradation. Was it a wayward child? Was it a miserable husband or wife? Was it financial circumstances? Was it war or famine? Was it a hateful boss or grandparents? The Lord said He would visit the sins of the parents on the children even unto the fourth generation. Was it ancestors who have caused the suffering? Tests with rats have shown that an incorrect diet can affect the posterity of the animal for as many as four generations. In all fairness to those who come after us, isn't it time we did something to help ourselves as well as our children by learning and living the law upon which good health is predicated?

The discovery of laws is the beginning of wisdom. To live by them brings perfection — a perfect knowledge without superstition or guess work. It may not be known except through revelation how the law works, but experience teaches that it works relative to what is done for the body. When we learn how to behave in regard to our health and refuse to live correctly by this understanding, we find ourselves in the category of the stupid. Gibby said:

> *Intelligence is the ability to discover and accurately interpret truth, coupled with the capacity to accept it and consistently apply it to daily living.*

There are those who always live in this category because they learn not by their mistakes and are either contemptuous of law and order or are so weak as to be persuaded and influenced by others who have convinced themselves without reason.

Society, leaders, family members, big money interests, and social pressures may encourage or impede health because the majority of people are like sheep — followers, like the sheep, unable to find their way alone. This is why Jesus likened man to sheep. The sheep is the only animal who by instinct cannot find its way home. Like sheep we need guidance, but we must still be the judge of what or whom we will follow. The inescapable fact is that the majority is always and has always been wrong and they become so, not on an individual basis, but because they follow the wrong leaders. Often the law is broken for the wrong reasons.

Gibby said:

> *Many of us live far below the character standards we are capable of achieving, simply because we fear the criticism or crave the approval of those among whom we work and live. But never allow such lack of approval to keep you from high resolve or noble deed, for even a fool does not respect a fool.*

As changes take place when a person begins to do even one thing in his favor, such as taking so simple a thing as slippery elm for an ulcer, the vast and wonderful possibilities reveal themselves if the whole law were understood and lived. Broken law exacts its final penalty like repentance. It does not take as long for nature to reward the body with health as it took to become sick. As soon as a beginning is made, the law becomes effective, working favorably in our behalf. When these laws are lived, restoration of health becomes an almost magical experience. When iridology is understood wonderful changes take place not only in our bodies but in our eyes. The iris is the nerve organ in the body which is the most sensitive and highly developed. It can instantly receive messages from any part of the body, reflecting pathology or restoration to health. The nervous system reflects the situation to the brain, which in turn causes the same situation to then be recorded on the eye.

Making a correct choice which proves to be predictable at all times develops strength; and if it is a side-step from the ebb and flow of humanity's erroneous ways, we find that we do not have to be so rigidly subservient to the way of the world. Each time we prove one of these choices to ourselves, whether it is confidence in answer to a prayer or overcoming a stomach ache by taking the right herb, we are a step closer to an intelligent independence of our own soul and body.

*Truth walks a straight and undeviating*
*path and shows partiality to no man.*

*Gibby 9/1/61*

Following are the testimonies of a few brave souls — among the many — who start and stop, slide back and go forward, in this process of self-discipline, who have persevered long enough to make some very interesting changes in their bodies and their irises.

**Tapeworm**

After attending a class taught by Mrs. Griffin I saw my eyes reflected a body in need of cleansing, so I started last June on the mild food diet and herbs. Immediately I became sick and stayed sick for six weeks when I decided to let up on it for a while. I didn't know how ill I was at the time. Some lumps appeared, they turned out to be malignant. I wasn't shattered, I knew other people had conquered the same malady through natural ways. I searched for every natural way I could find. I was one of the lucky ones. I had already started on the mild foods diet — being already familar with it made it easier. I had never quit taking the herbs. After the doctor's diagnosis I started again on the mild food diet.

After I had been on the diet (and herbs) about one month I passed some nodules with long black finger-like things on it. Odd looking, but the pathology report was negative, only mucus clumps. From that time to this (two months later) mucus has been passing out of my body. Then to my absolute horror, two days ago, I passed a tapeworm. That was enough for me, I wanted everything to shape up right now, but nature doesn't work that way. I can see by my eyes I have a way to go yet, before I have a completely clean intestinal tract. The path is before me, just wish I could travel it faster.

I love the Lord. He is my "guider". He has led me to the ways of truth and light. His is the credit and the glory and may he bless those who have the courage to stand, and especially those who stand and help others.

**Theora Sheets**

I would like to give my testimony to the effectiveness of LaDean Griffin's program of health. For the past three years I have been on mild food and the healing herbs that she describes in her book, "Is Any Sick Among You"? This has been a very, very choice experience for me and a very spiritual one because at the time that I began this program, I was very ill. The medical doctor — my gynecologist — had made tests with the pap smear and had indicated that I should have a hysteroctomy because there were so many places where all the tissue was so poor and the threat of cancer was there. There were — I don't know exactly how they count these things — but they talked about how the count was lymph high and that it was indicated that I should have a hysterectomy and do some testing. I went in at that time and did have a D & C and they came up with the thought that I should wait a little while and see how things went. At that time my husband was very ill with cancer and I didn't feel that I could spend time in the hospital with him so sick and I pleaded with the doctors to let me wait until he was better before I had a hysterectomy.

I also had many other health problems. I had had inner ear troubles for years, and I would lose my equilibrium. I would lose my hearing and would be very ill with that problem. I also had a very severe back problem. I was seeing a chiropractor periodically, and getting slight results because of this, but it was not a true healing. It was only a sort of a pacifying thing that would ease the pain in my back and leg. I was unable to even get in and out of the car without having to lift my leg in and out. I was losing control over the muscles in that leg.

I began this program and stayed on it faithfully using the mild foods and the herbs. Immediately, my health began to improve and I began feeling better. The healing crises have been disturbing and upsetting because you feel that you are going back into the sickness again as your body tends to throw off the waste. But I have waded through the healing crises and I have always come up feeling better after I have finished.

The doctors have made test, of course, since I have been very careful to get the tests every six months to check to see if there was any more cancer — any more problems there, and they have found nothing. In fact, two weeks ago when I went in for a checkup, the doctor said that he just couldn't believe how well my body has healed and how healthy the tissue is. I've also had heart problems in the past that have caused me lots of discomfort and have kept me from doing things that I enjoy doing and now at the present time I am working full-time. I'm very well. My heart never gives me any trouble and I don't have trouble with my back at all. My chiropractor has released me. He has shown me the X-rays that he has taken compared to the ones he took on me three years ago and he said, "I can't believe the improvement you have made in your bone structure and your bone quality." He said the "this shows that you are progressing in your health, rather than deteriorating." I am now age 55 and I feel better than I did when I was 35. My mind is clear and sharp and I am active and I love to be with people and share my feelings about this beautiful method of staying well and getting well. I might also say that I have had arthritis in the past, however, since I've been on the program it has gradually left me. I still have a few little nodules on my fingers, but I never have any pain from them and they are gradually getting smaller and going away. This is a beautiful program of health and it has not kept me away from the doctor's office. I have continued to be checked and to have the tests made that are necessary to know what is going on and I have been thrilled to see the improvement that I have made and the doctors have been thrilled with it also. I think that it is important to watch what is happening when you start on a program like this and continue to keep track of your health and improvement that is made. Good health is important to all of us, and it's important for us to know our bodies, to know the foods and herbs that are needed in our bodies to heal them when we're sick. I'm grateful for the knowledge that I have gained as I've worked on this program. It has been an inspiration to me to help me to improve my health and to live here upon this earth in joy and happiness and being able to help others. I am Theora Sheets and I'm happy to give my testimony to the truthfulness of this book and the great help that it is to those that read it and enjoy it and learn to live with what is given there as a way to find a better way of life.

**Gary Gillum** — picture page 50

"The Staff of Life Overcomes the Staph of Job"

"Mirror, Mirror on the Wall, Who's the fairest of them all?" quoth I. "Certainly not a 'it's-just-caused-by-adolescence' pimply, 'it's-not-your-fault-you're-not-Charles Atlas' skinny chassis such as yours," replied the reflection to me. With that outburst of subjective truth my frame fairly tensed with anger at my Jobian state, insomuch that I thought my bowel would straighten as a nail. (Sigh.) I guess I will have to outwait my adolescent years of pimplehood and awkwardness. But I am 24 years old now! And what do I have to show for it but horrid reminiscences of taxing dermatologists who froze, relaxed, drugged and wrapped my face, clearly silly, phisohexian hocus-pocus soaps and cold-then-hot-water-should-do methods to embattle my pores still further? (I was told chocolate caused my condition, but I didn't like chocolate.) What to show except for a missing appendix, whose 22 years were exhausted as a savior, along with staphylococcus bacteria, in releasing poisons from my warring frame. These conditions, of course, were caused by the medicians' brews in hospitals and campus health centers where the 'modern' priestcraft of medical malpractice flourished as did Baal in the days of Elijah. Three years following the mirror incidents, the lowly, mistreated, misunderstood but beautiful and delicious dandelion became an instrument in opening my eyes and set the stage for a meeting with LaDean. (Thenceforward my wife and I have annually celebrated that momentous occasion with wilted dandelions from our lawn.) It was not merely the word-of-mouth advertising which sent us to this remarkable woman, but the desire on the part of my wife and I to search for light and knowledge, especially since my graying skin condition dictated that we would try anything once to prevent my permanent mortification. After several visits with LaDean, she informed us that on my first visit she had felt that I might not make it. But I had followed her teaching as well as the promptings of the Spirit and soon found the way towards better health. (Even as I write these words I can feel miraculous changes within my body, changes which are quickening my body, my spirit, and my long-lost self-confidence.) In the beginning it was necessary to rid my body of those creatures who fed upon my innards parasitically, and as months passed I fondly referred to them as Max, Maxine and Maxmillions. Well, they have since departed this life with the aid of herbs, which possess one thing medicinal drugs lack — innate spirit. (Man cannot synthesize the spirit of an element or being, even if it is possible to duplicate physical and outward features.) A famous Mormon lawyer and Church leader, J. Reuben Clark, Jr. once speculated that the administration of the sick by the elders of the Church (James 5:14) is necessary in many cases of illness since the body's healing mechanisms had not learned all of the disease ridding functions while in its earlier estate. In addition, I personally believe that herbs provide an added intelligence to the body cells, training the individual cells and organs to slough off the poisons and bacteria and fight with renewed vigor. Not that the herb itself has no intrinsic medicinal value. On the contrary, its purpose is two-fold, for Peter the Apostle had not only ordered the crippled man to arise, but offered his own assistance by giving him a hand. An example might illustrate this: the psyllium seed which helped my colon slough off and eliminate an 18-inch tumor definitely assisted. (That tumor I should also have immortalized in glass and formaldehyde — only after I had allowed medical scientists to scrutinize it and wonder how I could have gotten it out except by the edge of a knife.) The cleaning of my pancreas by herbs was a most important aspect of my experience. But that process, even more so than that of ridding the body of parasites, depended equally on correct nutrition. The only 'nutrition' I had previously followed as the world's estimate of a balanced diet for a 'skinny toothpick' plus the opposite of a weight-reducing formula: a fatty formula, consisting of double portions of meat, fats and carbohydrates, three meals per day during my college years. Well, we all know that stuffing elephantine amounts of 'food' into an Ichabod Cranian body does nothing for the glandular imbalance of the system except worsen the condition and throw the body into a metabolic tailspin toward an early transition. Upon beginning a correct diet for my own body, I recognized that throughout my years of adolescence I had suffered from migraine headaches, lack of energy, boils, hemorrhoids and beginnings of stomach ulcers. It would take some time to alleviate entirely these sufferings. I lacked the sufficient energy in the beginning to go completely on a mucusless diet, so it was supplemented with goat's milk and plenty of baked potatoes, which I grew to love. Several years later, I find that my favorite foods are fruits, nuts and seeds, and these provide me with all the taste-bud enjoyment and energy I could ever have hoped for. True, I weigh no more, but the other diet had not solved the problem either. The important thing now is being rid of disease almost entirely. Decayed animal matter, cooked foods, and unassimilatable substances could hardly feed a compost pile, let alone a human body.

I could recite numerous incidents which prove to me that I have found light and truth in the way I now live, but I will let the reader rest after a few choice ones: 1) A two-week cherry fast last summer found tablespoons of mucus coming forth out of my sinuses, unaccompanied by any cold pain except the welcome pain of matter breaking loose in the head area. 2) While taking black walnut to extradite Max and his family of parasites, I felt the most angry rebellion ever in my lower bowel, and I felt it necessary to thank my large intestine, my heart and my other organs for the valiant battle they fought and won. 3) This past summer I experienced my first five-day lemon and honey fast. I hungered for no other foods and possessed an over-abundance of energy I previously had rarely — if ever — experienced. Utilizing the scientific method or reason is no precise method in determining the truth of anything, except temporarily. Only personal revelation and direction through the Spirit, providing the patient is teachable, is an eternal measure of eternal truths. Then the person will be able to discern his way through the temptations, quackeries, misconceptions and evil teachings of a world whose god is mammon, and whose goal is to imprison mankind in darkness. The result for anyone who perseveres will be a pure and holy temple fit for the permanent habitation of the Spirit while he yet lives, possessing not only a joyful, selfish life, but a happifying, selfless life of service to others. Who can think of a more wonderful blessing that that? I offer my thanks to she who offered her knowledge, wisdom and service in the face of blind persecution from those who are 'learned.'

The following herbs, along with a mucusless diet, enabled me to help my body (and will in the future help my body) to become healthier and clean itself of a pre-diabetic condition (cleansing of the pancreas), headaches, hemorrhoids and anemia:

Golden seal (¼ teaspoon three times per day)

Cayenne pepper (")

Lower bowel herb (")

Herbal pumpkin (four tablets per day)

Lipotrophic adjunct (two per day)

Additional herbs and vitamins depending upon condition, season and diet.

Today for a general cleansing of the body, which will someday bring me to perfect health I take the following herbs and vitamins:

Golden seal (¼ teaspoon twice daily)

Cayenne pepper (")

CS formula (")

Nervine (")

Herbal pumpkin (six tablets per day)

Vitamin E and vitamin $B_{12}$ (one each per day)

Lecithin granules

Additional herbs and vitamins depending upon diet and

condition. During the summer months I eat liberal amounts of sunflower seeds and exercise my neck twice daily in order to improve my eyesight. At present I need glasses only for driving. I used to wear them constantly.

Pumpkin seeds (as an aid to herbal pumpkin)

Fresh fruit (almost exclusively this summer.)

**Jeannine McKenzie** — picture page 50

I started on the mild foods in August, and I stayed on it very strictly without any deviation at all for about three and a half months without anything else. The only herbs that I took at this time were licorice and some nervine herbs once in a while but not regularly, and the thing that I noticed the most, was that previous to this, I had really had problems with my hips. I could hardly walk up and down the stairs or do anything after just two or three hours of even mild work in my home. After this period of time, I could work from early in the morning — five and six o'clock-till clear into the middle of the night without ever having this problem again. It never bothers me. This and just being able to get out of bed in the morning and not having that tired, tired feeling, which I'm sure came from the consumption of too much sugar and this type of thing. And how do I feel right now? I feel great! I'm taking other herbs and things now. I'm still on the mild foods, and I've added quite a lot of alfalfa, for my glands, and more of the nervine herbs because I realize that this is what I need, and as I go along, this nervous, uncontrollable urge to just get out and away from things and not do the things I know I'm supposed to, is gone. It's really been wonderful. I lost thirty pounds.

**Michael Tracy** — picture page 50

All too often, we tend to blame God for death, disease, and anything that goes wrong in our lives. This is, of course, both unfair and unreasonable. "It was never the intention, nor in the plan of God to produce disease, but as a logical consequence, thru disobedience of the divine laws of life, disease is produced." (The Mucusless Diet Healing System, Ehret, p.85.)

We are, then, largely responsible for our own physical well-being, and there is no doubt in my mind that each of us will be held accountable for the manner in which we take care of our bodies.

I believe now is the time to learn and live the true laws of health, endeavoring to recall from antiquity the precious knowledge that was once held sacred by our Biblical ancestors concerning these things. I believe that man can and will, with the help of the Lord, progress to the point where he will no longer be a slave to poor health. And I feel certain that these things will be accomplished with proper diet, herbs, and the work and inspiration of the Restored Priesthood.

Because of the Mild Food Diet, and the judicious use of herbs, my wife and I have taken long strides in improving our health. For us, this new, exciting way of life has given us a feeling of vibrant rejuvenation and well-being. We know that it has strengthened us spiritually as well as physically. We are grateful that we have learned these things while still young and we are certain that the children that will be born to us will reap many benefits because of our diligence.

Yes, there have been a few times when we wanted to "cheat" on the diet. More than once I have seen a steaming rib roast, a Dairy Queen, or a Cheese Whistle that looked appealing. None-the-less, the rewards received for avoiding these things are spectacular and well worth the efforts involved in cultivating self-discipline.

My wife and I look and feel younger. Our eyes have become brighter and clearer, and our strength and stamina have increased. We share a firm belief that our efforts are purposeful and in keeping with sound religious teachings. We have a strong testimony of the truthfulness of the Mild Food Diet, and we know that anyone who adheres to it will enjoy the same benefits and blessings that have come to us.

**Alexandra (Lexie) Veranth Harris**

Today, I am a happy, healthy, slim, 34 year old wife and mother of four children, counting our foster daughter. It was not always so!

It is my testimony that I have been to hell and back. I attribute my present happy state of affairs to prayer, the influence of the Holy Ghost testifying to me the truth of the diet as a higher law of eating so that I had the faith to stick it out when it took courage; and, to my loving, giving, understanding husband who would cook his own meat meals and supported me during my journey to health. He could so easily have made it harder on me or tried to make me feel guilty. At first I didn't have the willpower to cook and look at meat and bread dishes without eating them. I do now.

From infancy I have had what was diagnosed as chronic atrophic pyelonephritis. Which means I had one kidney stunted and atrophied from chronic infections. From two years before my first marriage and continuing through the ten years of its duration I was plagued with constant chronic kidney and bladder infections which at times caused such pain that I required hospitalization. During the rest of the time the infections were controlled by strong antibiotics and almost constant pain pills and narcotics from percodan to morphine. During my first marriage my husband poured literally thousands of dollars into doctor and hospital bills. And countless more into pills of all kinds. I had major surgery three times. Minor surgery four or five times. Hospitalizations excepting the above at least once a year. And I even sought psychological help because of my misery and inability to care for my family. I learned that anyone who is that sick with a family to care for and no hired help or consistent help is bound to go a little balmy upstairs. I spent quite a few hours depressed and many nights sobbing into my pillow begging our Father in Heaven to help me out of my seemingly inescapable pit, as the doctor's had said there was no known cure for chronic disease of my type

I bore three children, two of whom were born with hyaline membrane disease, 6 weeks prematurely. The youngest required hernia surgery at three months. They both suffered chronic severe constipation from toddler age until we went on the diet a year ago when they were seven and eight years old respectively.

Almost six years ago I was converted to the Church of Jesus Christ of Latter-day Saints; and that is when my life made a drastic and steady change for the better. I stopped drinking and smoking. Became pregnant with my third child who was born normally; but with bad colic. At that time I received my Patriarchal blessing in which I was cautioned to be careful of the things regarding my health so that I would have vigor of body and mind. This was a new concept for me as I had always gone faithfully to the doctor and been an obedient patient. Then my baby came down with diarrhea for five or more weeks straight and the doctor couldn't stop it. That's when I visited my first health food store and my blessing began to make more sense to me. I began to see the vision of health. I took a health and nutrition class and read everything Adelle Davis wrote and anything else I could get my hands on.

At this point I should mention that, at the time of my marriage I was a Seattle Seafair Princess and was blessed with a very nice figure. Well, disease, drugs, four pregnancies and my delicious, gourmet and pastry cooking had ruined it. I'm 5'6'' and with my bone structure should weigh around 114 pounds. During my years of sickness, I weighed from 130 to 145 pounds. After my third child, the first I had been able to nurse, I reasoned that since chocolate is bad for a nursing baby (I was a chocholic) and a milk drinking mother was good for one, I would kill two birds with one stone and eat ice cream. That would satisfy my sweet craving (a life long urge) and we'd all be happy. The day came I weighed 175 pounds and no end in sight. I was into Adelle Davis at the time and threw out all canned, packaged, and frozen foods; and ate a highly superior form of the great American diet. I began baking my own bread and switched to goats milk. My children still suffered from constipation which was a nightmare; but my weight over a period of two years gradually decreased until I stabilized at 145 pounds. I felt better because I looked better and was proud of myself. However, we were still sick, although not as drastically.

Then the children and I moved to Seattle. Now cooking to please myself, I cut out most meat and most sweets from the diet. At this time, I experienced my first great cleansing crisis, pneumonia; although at the time I didn't realize it. My dear mother nursed me to health and then we moved to Provo, Utah, where I had the good fortune to move in across the street from my future husband and next door to Gary and Lyn Gillum.

I had had the feeling deep down for several months that I should try vegetarianism; but the whole subject baffled and scared me. I had been raised closely associated with the medical profession both in my immediate family and as employers and had quite a knowledge of medicine both as a matter of curiosity and as a sympathetic viewer of the doctor's oftentimes selfless and dedicated role. And through my studies I had been scared by nutritionist's claims of the deleterious effects of strict vegetarianism. My curiosity and desire for truth led me to ask my neighbors about their way of eating as I had heard some pretty interesting remarks from acquaintances. The first thing they told me was about the delicious pizza they had made the night before with potatoes. All sorts of crazy scenes passed thru my mind. Then they acquainted me with LaDean's works. I read her book and knew by the burning and recognition of truth inside me that what LaDean wrote about was a higher law of eating. It made intellectual sense. It supplied the missing pieces in my jigsaw puzzle of knowledge about health and diet. I decided to try it; but put off following it strictly until after my impending honeymoon. I just couldn't see passing up all those fantastic Chinese restaurants and lobster dinners and fine pastries on my honeymoon. Well, you guessed! On the fifth day of my honeymoon (we took it with the kids) I ended up in Heber City Hospital hobbling from bathroom to emergency room and biting my lips in pain. From then on pain dictated my actions. I couldn't see repeating the endless expensive cycle of drugs and doctors with no cure in sight. So I threw away the hospital's pills and went onto fruits and vegetables, a few nuts and seeds, some raw juice and herbs for the bladder and kidneys. I took catnip for my nerves, golden seal and licorice root for my kidneys and for my craving for sweets, CS to cleanse my system, psyllium seed powder mixed with juice for my bowels (which were in horrible shape) and I fasted and prayed a lot. I slowly got better. I would have improved faster; but, I had a hangup about taking enemas. One day after coming off a fourteen day fast my tummy swelled like I was six months pregnant within three days after I resumed eating. I forced myself to take an enema and the pain and swelling subsided within the hour. I discovered that poisons will pass into the bladder from a toxic colon. I have since been concentrating on strengthening my bowels. I now consider myself healed of my main pain causing problems; but have a way to go, before completely healing all of the organs and areas of my body that were almost destroyed by drugs and faulty diet. I have my teenage figure back and look about ten to fifteen years younger than I did at my worst. People guess my age to be about 15 years. During the period of my greatest cleansing I appeared skinny and haggard to others, even though I had progressed far enough to have conquered the pain. But continuing the same diet of fruits, vegetables, nuts, seeds, a small amount of raw honey and juices I gradually gained back a small amount of weight to what is normal for me and gained energy and color. The only dietary change I underwent was to switch from cooked to raw vegetables, shredded and added to salads and an increase in herbs. And more regular enemas, of course. I also now try to drink at least two quarts of carrot and other fresh raw fruit and vegetable juices each day to build up my body nutrition. I've thrown out most of my supplements, retaining vitamins C and E, and maybe garlic pills. I used to spend a fortune on food supplements after I threw out the prescribed medicine. Now I use juices and herbs to cleanse and heal. Lately I have used the skinny formula, which has produced the most terrific head cleansing and pouring forth of mucus I would not have believed possible. I have had iridology pictures taken; but my body was so deteriorated that I saw only slight improvement. If I had had recent pictures however you would notice that my eyes have turned from what people saw as a darkish grey but mistook often for brown, to a much lighter blue with heavy white acute healing lines. In every area where there was a black closed lesion there is a thick mass of healing lines.

During my illness, and cleansing healing change of diet I made the mistake of worrying my family long distance by telling them what I was and wasn't eating and how wonderfully skinny I was getting. Then six months ago we were able to visit. You can imagine my great feeling of exultation as I saw their eyes light up in delighted surprise and heard their remarks of approval and curiosity. Then just after Christmas my mom, Ruth Veranth, called and asked me to visit and teach her the diet. I went armed with LaDean's books, some herbal formulas, an iridology chart and the determination to buy her a juicer. I bear my testimony that our Father answered my prayers and if the most famous credentialed doctor in the world told me I was wrong or if the powers that be all rallied against it I would not change my way of eating and herbing, as my children and I suffered hell with our bodies and now know budding health and have happiness.

The following is a poem my mother woke up with on her mind one morning while I was teaching her the program.

Here's to tricks in '76

No more will the bod be a jinx.
'Twill look and feel like a merry minx
Joy and laughter will erase the kinks,
Charles and I out to the links,
And no more garbage to make us finks,
Eat it and the whole world shrinks
And a house full of junk saps the energy methinks.
To the depths of despair the weary one sinks.
Careful now —trim your wicks
Here's to celestial heights in '76!

**Ruth Veranth**

It is with pure joy that I offer my testimonial on the efficacy of LaDean's nutritional "way of life." Since my daughter, Lexie, brought me the knowledge six weeks ago I've shed 20 pounds, and have never felt more buoyant, energetic, free and alive than with my new found delicious diet of live fruit, vegetables, herbs, nuts and seeds. Dramatically, unhealthy cravings and leg muscle cramps disappeared as if by magic. I want to shout my praises to the housetops; my friends are amazed and ask for this pragmatic answer to one of life's most important problems-how to eat right and greet each new day free — with sparkling enthusiasm, brimming over with energy and happiness.

**Wyonna Kruse** — picture page 50

Something was wrong! I was so tired all the time, with a shaky feeling inside. At times my body appeared to be going 100 miles an hour, and I felt I had to get food inside of me to slow this down. This extreme tired feeling persisted and I kept tuning it out. My husband was experiencing extreme physical problems, and I knew this was not the time or place for me to be getting sick. At the insistence of a friend, I went to a doctor.

The usual tests and x-rays were done. I took huge amounts of protein to build my adrenalin. I began swelling and I experienced so much bloating when I ate, that I had to take chewable enzymes. My daily intake of hormone pills and thyroid was astounding. I had learned to check my vitamins for purity from cold tar, and I knew they were not causing my discomfort. (Incidentally, should you be interested in your own vitamins, place the vitamin in an old spoon, heat from a lighted candle turns it black, if they are made from cold tar. Vitamins showing this coloring should be dumped.)

Time passed. I went again to the doctor. I showed a very high glucose test shooting up to 480 and then dropping to 47. My doctor showed me the blood tests under the microscope. When he pointed out the little globules moving so sparkly in my bloodstream, I asked him, "Are these cancer cells?" He evaded my question indirectly by saying, "You do have a bad infection in your blood." He gave me a prescription for several drugs and one potent anti-biotic. None of these drugs appeared to be helping my condition. My doctor made a serum from my urine, which I took twice each week. Nothing helped. Then I heard about the book, *Is Any Sick Among You?*

I began attending the classes on iridology. When I saw my own eyes, I realized that I was much worse off than my husband, who had suffered a coronary. I knew that I had a strong inherent body and with the help of the Lord, and my family and friends, I could discipline myself to get on this cleansing program and stay on it.

Those who have been on the program know that when you first begin the routine, you clean so fast that the toxic waste moves into your bloodstream and makes you deathly ill.

I hadn't eaten much this day and about dinnertime I began noticing terrible pains. I had committed myself to attend a meeting this evening. While sitting in this meeting I experienced such terrible pains, severe like labor pains only they were in my liver and pancreas regions. I could time their occurrence and reoccurrence just like labor pains. I was almost out of my mind with pain. I had to be excused from the meeting. I didn't want to call the doctor and I had to get relief, so I took a coffee enema. (Coffee stimulates the liver) I took a quart of water, but I couldn't expel it. Nothing would come out. For 30 minutes I massaged and worked my colon. I rolled on the floor, I walked, I tried different positions. The pain was still intense, but not as intense as before I put the water in. I took another quart of the water and it was at least 15 minutes later that I was able to expel the water with a large clump or mass. I should have had this mass checked. I'll always believe that it was a tumor that had passed from my body.

Herbs will loosen and pull the waste from the intestines. The intestines are like a sink, they can become plugged up. We wonder how they can become so plugged up. This is what had been happening to me. I had been pulling all this waste matter from my intestines which caused this intense pain. People who have been on laetrile (Apricot seed) and chaparral herb, over a period of time, will experience this intense pulling as the fecal loosens from the body. The pulling becomes so great that it appears to cause a bloated feeling and you wish you could take a knife and "stick" yourself as the farmers did their cows in the years back.

My purpose in taking time to write how I feel physically, is for you readers to realize the urgency to get on this program, if you are having problems. I was thrilled with my doctor's last report. My protein was higher than it has ever been and my hypoglycemia is much better.

I have noticed that people will give up anything but their food, and the food habits they enjoy. Usually it takes a real live "jolt", mine was "terminal disease" to push one on the program. Self-discipline is the key to the program, through the years we rely so much on the arm of the flesh. I keep telling myself that I have to stay on the program — plan certain days for the cleansing process with herbs and the follow-up with a strict diet of mild foods. The toxic in my body has to be released for me to get completely well. I have had to learn to live each day as the sun comes up. Some days my mind wants to go, but I have difficulty dragging my body along. Without my work, I'd give up. The Lord works in a mysterious way — maybe this is the reason he is teaching me to discipline myself. I have learned to say, "I'm not on a diet, I have low blood sugar and I eat mild foods." I am feeling much better now than I have felt for so long that I can't believe one can feel this good and still be not well. I didn't get this disease overnight. My body has probably been working on this for 30 years.

Don't let anyone tell you this is an easy program to follow. There are good days and days that are bad, but as Abraham Lincoln states so beautifully, "God is the silent partner in all great enterprises," . . . and I have gained an inner strength and peace of mind by placing my burdens on Him.

From the pen of the poet, Phillips Brooks, "O my young friends, the world is beautiful and . . . life is full of promise."

Wouldn't I be robbing God if I were to give less than my very best to the program?

# CHAPTER FIVE

# TO STAND ALONE

The Lord has told us to become as a little child in order to enter the Kingdom of Heaven. The trust and love of a child, even in a parent not worthy of that trust, is a beautiful thing. Is this the child-like attribute He was talking about? Can a child-like trust in God be represented continually as a way of life? Too much time is spent in trusting in the arm of flesh and not enough in trusting God. We lean so heavily upon each other that we miss the opportunity of direct contact with a higher power. The conflict of trusting in the arm of flesh when everyone is so self-seeking, leaves little room for inner rebuilding of the soul. Unique individuality and intense spiritual power must first be achieved alone, away from the crutch and weaknesses of other people.

Maturation depends upon mother and dad; but even though we then begin to fly alone we still need guideposts to follow. Because we have had little training in these new pursuits, we either muddle along making all the mistakes humanity is capable of making, or we wisely seek the eternal wisdom of the scriptures and great minds of the past to discover what standards were upheld in their way of life. As these guideposts dot the landscape of our world, eternal trust is discovered. It then takes form and correct pattern is recognized, a definite procedure that must be observed in order to find man's most sought-after prize — joy. Joy follows good health, but in order to have good health we must use wisdom. To acquire wisdom, we must learn either by trial and error or by studying the past mistakes of others.

As the complex problems of making a living and raising a family begin to crowd in upon us, we are often persuaded and influenced from all sides to make the incorrect decision. Before we can stand with God we must be able to stand alone, away from the opinions of others, away from leaning on other people. This is what is called courage. Would we be happy with our adult children if they came home every day for us to tie their shoes, or to ask us to do any of the things an adult is supposed to be capable of doing alone? By the same token, there are things we must learn to do by way of making certain judgments and choices which would be as bothersome to the Lord as if we had asked him to come down and tie our shoes everyday. Some people have the idea that they must ask God to make all of their decisions for them and they call this faith. Somewhere along the way we must learn what God's laws require and make plans to stay within those boundaries. When I think of how honest and just a man my father was, I remember before I did anything questionable or wrong that I would ask what my dad would have done or thought.

When my dad was asked to take a high-paying job as a radio announcer and refused the position, I said to him, "I'll bet I know why you didn't take the job. You have always said radio announcers were liars, advertising obnoxious products detrimental to health." He said, "That's right, I could not advertise cigarettes, beer or even foodless foods sprinkled with sugar, and tell children these foods would make them strong." Everyone has not had the example of goodly parents but we **do** have the example of great minds, prophets and scriptures to light our path.

There is a great difference between being completely dependent on the Lord, completely subservient, completely loving and completely child like than not being able to make any decisive choice or sacrifice. The Lord placed a head upon our shoulders with a body attached and forced us to do only one thing: **choose.** It has been regally said by Edward Young, the English poet of the sixteenth century,

> On the soft bed of luxury most
> kingdoms have expired.

If God made all of our decisions we would always make the right decision. Therefore we would be like a robot, without learning in the final analysis, why God's way is best. The Talmund makes a magnificent observation.

The best preacher is the heart,
the best teacher is time,
the best book is the world,
the best friend is God.

To teach us the truth about pleasure and pain, Pope the English author said,

Reason's whole pleasure,
all the joys of sense,
lie in the words
health, peace, and competence.

What great rewards there are in accomplishment of making the right choice and of doing something well. Robert Louis Stevenson said,

He has achieved success who has lived well,
laughed often and loved much; who has gained
the respect of intelligent men and the love
of little children; who has filled his niche
and accomplished his task; who has left the
world better than he found it, whether by an
improved poppy, a perfect poem, or a rescued soul; who
has never lacked appreciation of
earth's beauties nor failed to express it; who has always
looked for the best in others
and given the best he has; whose life is
an inspiration, whose memory, a benediction.

Real joy is found in obedience to the law in common and ordinary things. A mother's domain is a palace filled with ordinary things where all the normal necessary desires are gratified: the table, the bath, a clean, comfortable bed, the laundry. The beauty and refuge from the world is the home, which she creates for her family. S. J. Hale:

We need not power or splendor,
Wide halls or lordly dome!
The good, the true, the tender,
These form the wealth of home.

For the father, joy is also found in the ordinary. Lord Houghton said,

> Do not grasp at the starts, but do
> Life's plain common work, as it becomes
> certain that daily duties and daily
> bread are the sweetest things of life.

Plutarch, scholarly historian of the Greeks, said,

> That state of life is most happy where
> superfluities are not required and
> necessities are not wanting.

The satisfaction of winning the security for his family by the sweat of his brow — knowing he is the provider, teacher and protector of his family — brings joy and splendor to manhood. Plutarch reflects in his historical writings,

> The man who first brought ruin upon the
> Roman people was he who pampered them
> by largesses and amusements.

Seneca, the Roman Statesman, said,

> Great is he who enjoys his earthenware
> as if it were plate, and not less great
> is the man to whom all plate is not more
> than earthenware.

Rousseau, the French Philosopher, said,

> Temperance and labor are the two best
> physicians of man; labor sharpens the
> appetite and temperance prevents from
> indulging to excess.

Socrates, wisest of the Greeks, said,

> Contentment is natural wealth,
> luxury, artificial poverty.

Great men of the past have recorded their findings in decisive language to light our path if we but search and try to find. Cervantes defined proverbs as short sentences drawn from long experience. These philosophies, with their old familiar road signs, seem to appear in all languages and for all people as the infallible laws to happiness. Horace Greely said,

> The darkest hour in any man's life is when
> he sits down to plan how to get money
> without earning it.

J. M. Gibby said,

> Iniquity creepeth in wherever selfishness,
> pride and haughtiness unlock the door.

Solomon taught us,

> Go to the art, thou sluggard;
> Consider her ways and be wise;
> Which having no guide, overseer
> or ruler, provideth her meat in
> summer, and gathereth her food
> in the harvest.

My dad used to say, but I do not know who wrote it,

> It is following the path of least
> resistance that makes rivers and
> men crooked.

Goethe the poet said,

> If you want to make life easy
> make it hard.

Shakespeare said,

> There can be no better sign of a
> brave mind than a hard hand.

To conquer self is a part of the pathway to joy. DaVinci said,

You will never have a greater or lesser
dominion than that over yourself.

Aristotle said,

He overcomes a stout enemy
who overcomes his own anger.

Solomon said,

Who then is free? The wise man
who can govern himself.

Christ laid down the fundamental principles of self-discipline as He ended His Sermon on the Mount,

Enter ye in at the strait gate,
for wide is the gate and broad
is the way that leadeth to destruction
and many there be which go in thereat;
because strait is the gate and narrow
the way which leadeth unto life,
and few there be that find it.

When we find great, happy and successful men, we find men of noble purpose and high spirituality who follow a straight line to accomplishing a great task by setting righteous goals and aiming courageously toward them. Then when we seek to find health by living the rules we have brilliant words to light our path. Paul said,

Be not deceived; God is not mocked, for
whatsoever a man soweth that shall he
also reap. He that soweth to the flesh
shall of the flesh reap corruption. He
that soweth to the spirit shall of the spirit
reap life everlasting.

Byron said,

Vice digs her own voluptuous tomb.

Channy said,

Sensuality is the grave of the soul.

Pliny said,

Lust is an enemy to the purse,
a foe to the person, canker to the mind,
a corrosive to the conscience, a weakness
of the wit, a besotter of the senses, and
finally a mortal bane to all the body.

Longfellow said,

The blossoms of passion, gay and luxuriant
flowers, are bright and full of fragrance,
but they beguile us and lead us astray and
their odor is deadly.

Socrates said,

The end of life is to be like God and
the soul following God will be like him.

Daniel Webster said,

A conscience void of offense before God
and man is an inheritance for eternity.

J. M. Gibby said,

He achieves in vain who in the process
does not develop sound character, only
he who is constantly learning and serving
has a truly abundant life.

The Psalmist wrote,

What is man, oh Lord, that thou art
mindful of him, and the Son of man that thou
visitest him. For thou hast made him a little

> lower than the angels and hast crowned him
> with glory and honor. Thou madest him to have
> dominion over the works of thy hands; Thou
> hast put all things under his feet. Oh
> Lord, our Lord, how excellent is thy name
> in all the earth.

We need to rejoice that we have been fashioned so beautifully so as to have happiness and success through obedience to the law. Samuel said,

> Behold, to obey is better than sacrifice
> and to hearken than the fat of rams.

The beautiful words of the Lord to Joshua as he led his people into the land of promise, rings loud and clear — giving us hope and courage to go on:

> Only be thou strong and of good courage; be
> not afraid, neither be thou dismayed; for the
> eternal is with thee, whithersoever thou goeth.

I have touched only upon a few of the great writings of the past, with the hope that our horizons may have broadened, that we have emerged from their words, realizing that life is a beautiful experience and that when we seek to watch for the road signs — yet standing on our own feet building ever building courage, goodness and beauty into our lives — we will not be easily turned aside by evil and designing men in the last days. We will not allow someone else to make such serious decisions concerning our health nor to our freedom of choice such as.

Some so-called experts in the field of medicine, who are still standing on the intellectual premise that the next few years will be an immunological period are even looking forward to transplanting genes and scientifically make chemical changes in the molecule. They think they can stop manic depressives and women outnumbering men. They call it genetic engineering but I call it "trying to play God." They fail to understand the principle that after curing a person of his mental or psychological problems he seems to be a carrier to his offspring and so doctors intend to immunize against such defects. Immunology defined is: the total means by which the organism defends itself against all foreign substance. Are we going to allow a few scientists to run us and our children on this gauntlet of rearranging our genes so they can swell with pride and say "I made him what he is" — like a robot or a frankenstein monster. Isn't it time we began to make these individual changes in our own lives and in our society? Isn't it time we recognized that science has far too long used us as the scapegoat in their endless pursuit to find answers so as to change the existing laws of heaven and earth rather than seeking to live by the laws that already exist? Isn't it time we improved upon these laws, living them to perfection, rather than trying to change them to suit someone's idea of how God should have made his world? We all know that Great Art re-creates realisms in a perfect form. Why not make realism into a perfect form rather than like the art of Picasso and Rodin who have tried to change beauty into their own grotesque, senseless idea of what art should be.

God has told us:

> Be ye therefore perfect even as your
> Father in heaven is perfect.

Science has far too long looked for the answers in the wrong places when it comes to human life. Besides who is going to play God and pick and choose our genes — only these unsure, imperfect men who feed their egotistical minds with their own supremacy and intellectual superiority. Are we raising more Hitlers or are we merely worshipping science in the same way men of old worshipped the golden calf?

Seven hundred years ago the church ignored Roger Bacon's plea for restraint. The myopia of the people who lived with superstition caused Roger Bacon to cry out for scientific study. He feared that their reliance on the church would keep them ignorant. We have another kind of myopia in our day and time when man has developed a blind, single-minded reliance, placing all his hope and future with science and so-called men of learning. Isn't it time for us to re-evaluate the "old" laws of God and apply them today to our environment and our relationships with one another? Or would we prefer to become perfect people robots at the hands of a few so-called intellectuals?

The Scientist and Doctor have become the great High Priest —

where medicine has become a religion. But like the sixteenth century church, medical science is going to have to brace itself for the reformation, because thinking people everywhere have become disenchanted and are looking for and finding alternative solutions to their problems. They are learning preventive medicine and discovering that they do not want this dictatorship protection of a science which holds a pistol to their head with the continual claim that it is good for them. Neither does modern man any more want to be baby-sat by a dangerous maniac. Concerned men and women everywhere are going back to the old simple rules which God has set down and finding health and peace, as they learn to stand on their own feet, take care of their own bodies and in effect refuse to lean on the arm of flesh. There is great courage shown in the quiet consistence of well-doing.

# CHAPTER SIX

# TO WALK BY FAITH

It has often been said, "The eyes are the windows of the soul." In these past chapters we observed that the eyes are the windows of the body, but we must not forget that they are also a reflection of the soul or spirit. Personologists maintain that certain physical features of the eyes reflect certain personality traits. It is interesting that all feelings and emotions (according to personology) are located in and around the eye structures; for example, a person who is tolerant has a wide space between the eyes, an affectionate person has a large iris. Analytical ability shows on the eye lids and a critical person has a downward slant to the outside of the outer canthus. How the eye is positioned in the socket tell us something else. When the eye is deep-set, a person is thoughtful with many serious ideas; when it is not deep-set the person has great drive. Humor is shown by the crows-feet at the corners of the eyes. High-curved brows indicate dramatic flare; straight eyebrows show aesthetic traits. When the brows are straight and meet in the center, we have the artist. Sharp angular eyebrows show a good measurement ability, perception in mechanical judgment. When there is a good ability to use words the texture of the skin on the lower lids has pebble-like or criss-cross lines rather than smoothness. Good judgment shows up in the placement of the eyes, where the eyes are equally placed, with one neither higher nor lower than the other.

Personologists also maintain that these features can change if the trait is recognized and there is sufficient desire to change. This desire stimulates the already computed habits to make a correction in personality. Like learning to drive it takes continual conscious effort to make this change. As the change is occurring, the facial features reflect the trait being changed and adjust accordingly. Have you ever observed an elderly person who has lived a saintly life, whose beauty and softness seem to defy all description? Then have you also observed the person who has lived a life of sham, gradually developing an ugliness of soul, where features change, transforming not only physical features but reflecting the evil from within, out through the eyes? Evil or good reaches out from the windows of the soul, and people quickly respond.

An evil look in the eyes, an empty look, a lonely look, a lost look, a weird look, an insane look, a weary look, a look of love, a look of pain, a kindly look, a look of gratitude, a generous look, a forgiving look, a tender look, an embarrassed look, a look of fear, a look of remorse, a look of joy, a look of grief or anger: all of these and more can be seen instantly, reflecting outward from the eyes, from the soul and without words. These reflections are responded to by anyone within range, including our quiet, dumb friends of the animal world.

This reminds me of the time our big male collie dog was patiently allowing our recently-weaned kitten to snuggle up to him and suck. As I came out the back door and saw with surprise what was going on I said in a voice that could only have shamed him, "Pepper, for goodness sake, what are you doing?" He quickly shook them free, gave me an embarrassed look and walked slowly away with head down and tail between his legs. "The eyes are truly the windows of the soul." Soul to soul through the eyes is one of the best means of communication with which we have been blessed.

It is with interest that we note the discoveries of great men who have pioneered the Bates Method of eye exercise, who have learned that the shape of the eyeball is controlled by external muscles, which respond instantaneously to action, and that no refraction state, normal or abnormal, can be permanent. Consequently there are many things that enter into the ability to see well, and not all of these interferences with accurate sight are by any means identified strictly with the body but must, by all logical reasoning, be associated also with the spirit within the body. You have heard it said, "I was so mad I couldn't see straight." Whether we realize it or not, all the things we feel, think and do have a definite effect on what we project out through the eyes as well as the color beauty of our auras.

I found a small scrap of paper among my dad's papers in his own handwriting which said,

> May we always search for the good and
> beautiful, in living and good books,
> so that we might be better equipt to be
> of service to others.

A thing of beauty is considered to be a joy forever. There is nothing more beautiful than the reflection from the eyes and the aura of a great spirit, filled with an enthusiastic love of living and people.

If we would realize how often we are not a picture of beauty and not radiating forth from our inner soul those lovely things that bless and bring joy to the beholder, we might learn to think more on a positive and righteous level.

Carlyle has left us a forceful paragraph to help us in our searching, that we might reach into more lofty realms

> He is of the earth, but his thoughts are with
> the stars. Mean and petty his wants and desires;
> yet they serve a soul exhalted with glorious
> aims — with immortal longings — with thoughts which
> sweep the heavens and wander through eternity.
> A pigmy standing on the crest of this small
> planet, his far-reaching spirit stretches
> outward to the infinite, there alone finds
> rest.

Puny man in his reaches heavenward seems so small in the vast space of time and eternity, and yet Shakespeare wrote,

> What a piece of work is man; How noble
> in reason! How infinite in faculties!
> In form and moving, how express and
> admirable! In action how like an angel!In apprehension,
> how like God!

Emerson gave us a lovely thought,

> As much of heaven is visible as we
> have eyes to see.

It is not how much we can see but rather what can we contribute to beauty in the sight of others that is important. How can we be a light in the darkness so that this darkness might flee from us as well?

Helen Keller, who lived in darkness, has beautifully said,

> There is no lovelier way to thank God
> for your sight than by giving a helping
> hand to someone in the dark.

When we have learned a truth do we share it? When we have had a spiritual testimony do we hide it away and forget it? Are we like the ship's captain who, in a bad storm, began to pray, "Dear God, I haven't called upon you for fifteen years and if you will save me now I won't bother you for another fifteen years." Can we learn to pray always, not in a bothersome way, but rather with gratitude and hope? Can we stand on our own feet and yet trust that all things are done for our best good? Can we learn the laws of God and seek to obey them without being bitter about those laws, wishing it was some other way we might devise? The apostle Paul said,

> Love worketh no ill to his neighbour:
> Therefore love is the fullfilling of the law.

Love is the beginning of faith, the reaching out in trust to that God who created all the wonder and glory that is life on earth. Mark remembers a promise of the Master,

> Therefore, I say unto you what things so
> ever ye desire, when ye pray, believe that
> ye receive them, and ye shall have them.

In order to be in the company of God we must desire to be like Him, to obey Him in all things. Paul gave some excellent instructions to his brethren on how to become more God-like,

> Finally, brethren

> Whatsoever things are true,
> Whatsoever things are honest,
> Whatsoever things are just,
> Whatsoever things are pure,
> Whatsoever things are lovely,
> Whatsoever things are of good report,
> if there be any virtue and if there be any praise —
> think on these things.

Christ said,

> Behold, I stand at the door and knock;
> if any man hear my voice and open the door,
> I will come in unto him and sup with him
> and he with me.

A spark from a fire will only rise to the extent that it reduces itself to flame — if there is any dirt on it, it will rise only feebly and the dirt on it will soon consume its flame and bring it back to the ground. Not unlike the flame is man. With the weight of dirt and dross clinging to him, his soul will rise upward only feebly.

If we wish to rise we must seek the good, the wise, the pure, the gentle, the kindly, the loving, the beautiful. Lincoln said,

> Die when I may, I want it said of me by
> those who knew me best, that I always plucked
> a thistle and planted a flower wherever I
> thought a flower would grow.

In order to walk in faith and listen to that still small voice that whispers "all is well, you are here to learn my child and you are learning well, fear not for I walk beside you with love and tender concern" — we must live with lofty thoughts, high resolves, noble deeds, kind and loving thoughts, seeking after that which is good and pure light. Walt Whitman said,

> Keep your face toward the sunshine and the
> shadows will fall behind you.

There is much in this world that is shoddy, filled with degradation, pain and sorrow because people seek after evil and in its discovery, reap the wrath of a just God upon their heads. There is a great difference between the wrath of God and the refiner's fire, and when we begin to live by the spiritual rather than by the sensual, we soon discover the difference. Paul understood the difference very well when he said:

> By whom also we have access by faith into this
> grace wherein we stand, and rejoice in the
> hope and glory of God. And now only so, but we
> glory in tribulation also; knowing that tribulation worketh patience; and patience experience; and experience hope; and hope maketh not ashamed; because the love of God is shed abroad in our hearts by the Holy Ghost which is given unto us.

May the love of life and God fill you with the beauty of spirit which sheds its light forth from your eyes and the glow of your aura — to enriching the lives of all you meet — is my prayer for you, that such lights may begin to glow in the darkness and fill the whole world with the warmth of heavenly light. May you have eternal part in such glory.

# BIBLIOGRAPHY

**Books:**

Bates, W. H. *Better Eyesight Without Glasses*. New York: Holt, Rinehart & Winston, Inc.

Bealle, Morris, *Dangerous Doses*. Washington, D. C.: Columbia Publishing Company.

Bealle, Morris. *Super Drug Story*. Washington, D. C.: Columbia Publishing Company.

Best and Taylor. *Physiological Basis of Medical Practice*.

Boyd. *Textbook of Pathology*.

Breasted, J. H. *Edwin Smith Surgical Papers*. Chicago: University of Chicago, 1930.

Christopher, Dr. *Herbal Manuscript*. Provo, Utah: Bi-World Publishing.

Coon, Nelson. *Using Wayside Plants*. New York: Hearthside Press, Inc.

Culpeper. *Culpeper's Complete Herbal*. London: W. Foulham & Company, Ltd.

Dawson. *Medicine in Ancient Egypt: Annals of Medical History*. Vol. VI. Encyclopaedia Britannica, Library Research Serice, 1924.

Ehret, Arnold. *Mucusless Diet Healing System*. Beaumont, California: Ehret Lit. Publishing Company.

Forbes, Allen. *Our Garden Friends the Bugs*. New York: Exposition Press.

Gibby, J. Melvin. *Sand for the Rails*. Salt Lake City, Utah: Deseret Press.

Griffin, LaDean. *Herbs to the Rescue*. Provo, Utah: Bi-World Publishing.

Griffin, LaDean. *No Side Effects: The Return to Herbal Medicine*. Provo, Utah: Bi-World Publishing, 1975.

*Holy Bible, The*.

Howell. *Textbook of Physiology*.

Gytcgebs, Alma R. *Indian Herbology of North America*. Ontario, Canada: Marco.

Illich, Ivan, Calder and Boyer. *Medical Menesis*.

"Intestinal Parasites." *Modern Medical Councelor*, Chapter 42. Mt. View, California: Pacific Press Publishing Association.

Jensen, Bernard. *The Science and Practiče of Iridology*. Solano Beach, California: Jensen Products and Publishing.

Kadans, Joseph. *Encyclopedia Fruits, Vegetables, Nuts and Seeds for Healthful Living*. West Nyack, New York: Parker Publishing Company, Inc.

Kadans, Joseph. *Modern Encyclopedia of Herbs*. West Nyack, New York: Parker Publishing Company, Inc.

Kirschner, H. E. *Nature's Healing Grasses*. Riverside, California: H. C. White Publications.

Kloss, Jethro. *Back to Eden*. Coalmount, Tennessee: Longview Publishing House.

Kreig, Margret. *Green Medicine*.

Kreigg, Theodor. *Fundamental Basis of Iris Diagnosis*.

Kritzer, J. Haskel. *Testbook of Iridiagnosis*. Chicago: 1924.

Livingston, Virginia. *Cancer: A New Breakthrough*.

Loewenfeld, Claire. *Herb Gardening*. London: Faber and Faber.
Lucus, Richard. *Nature's Medicines*. New York: Awards Books.
*Master Herbology*. Ogden, Utah: Research Technical Service.
McCoy, Frank. *Fast Way to Health*. Los Angeles: McCoy Publishing Inc.
Meyer, Joseph. *The Herbalist*. Brand McNally Company — Conkey Division.
Netter, F. *Volume of Medical Illustrations, Nervous system*.
Shohl. *Mineral Metabolism*.
Steiger, Brad. *Medicine Power*. Garden City, New York: Doubleday.
Taylor, Thomas. *Porphyry: On Abstinence from Animal Food*. Barnes and Noble, 1965.
Tobe, John. *Proven Herbal Remedies*. Canada: Provoker Press.
Vogel, Virgil J. *American Indian Medicine*. New York: Ballantine Books.
Wigmore, Ann. *Why Suffer*. Boston: Rising Sun Christianity.
White, Ellen G. *The Ministry of Healing*. Mt. View, California: Pacific Press Publishing Association.
Young. *Handbook of Roentgen Diagnosis*.

**Periodicals:**

"DMSO Results Fantastic," Pleasanton, California, *Herald* and *News*, December 20, 1969.
Fredricks, Carlton. "Hotline to Health," *Prevention Magazine*, June, 1974.
Helleboe, Herman E. "Recognition for Need to Prevent Sickness," *American Journal of Public Health*, May, 1971.
Knowles, John H. "The Coming Change in Medicine," *Intellectual Digest*, February, 1974.
Merrill, Joseph E. "Eat Flesh Sparingly," Address in Tabernacle, April 5, 1948.
Mooney, Booth, "American people are eating more chemicals per capita than ever before,"*Plain Truth*, February, 1970.
*National Observer*,November 8, 1972.
"Needless Surgery," *National Observer*, July 29, 1972.
Smith, Tony. *London Times*, March 14, 1975.
Stevens, Guy P. "Some Foods Are Drugged," *The Improvement Era*, March, 1954.
Sullivan, Bernice. "The Organic Conversion," San Francisco Examiner and Chronicle.

# INDEX

*(Quotations by famous men are found throughout the book –their names are followed by a +.)*
*(Herbs are listed in CAPS; major sections marked in bold.)*

**T**



**U**



**V**

**W**



**Y**



**Z**

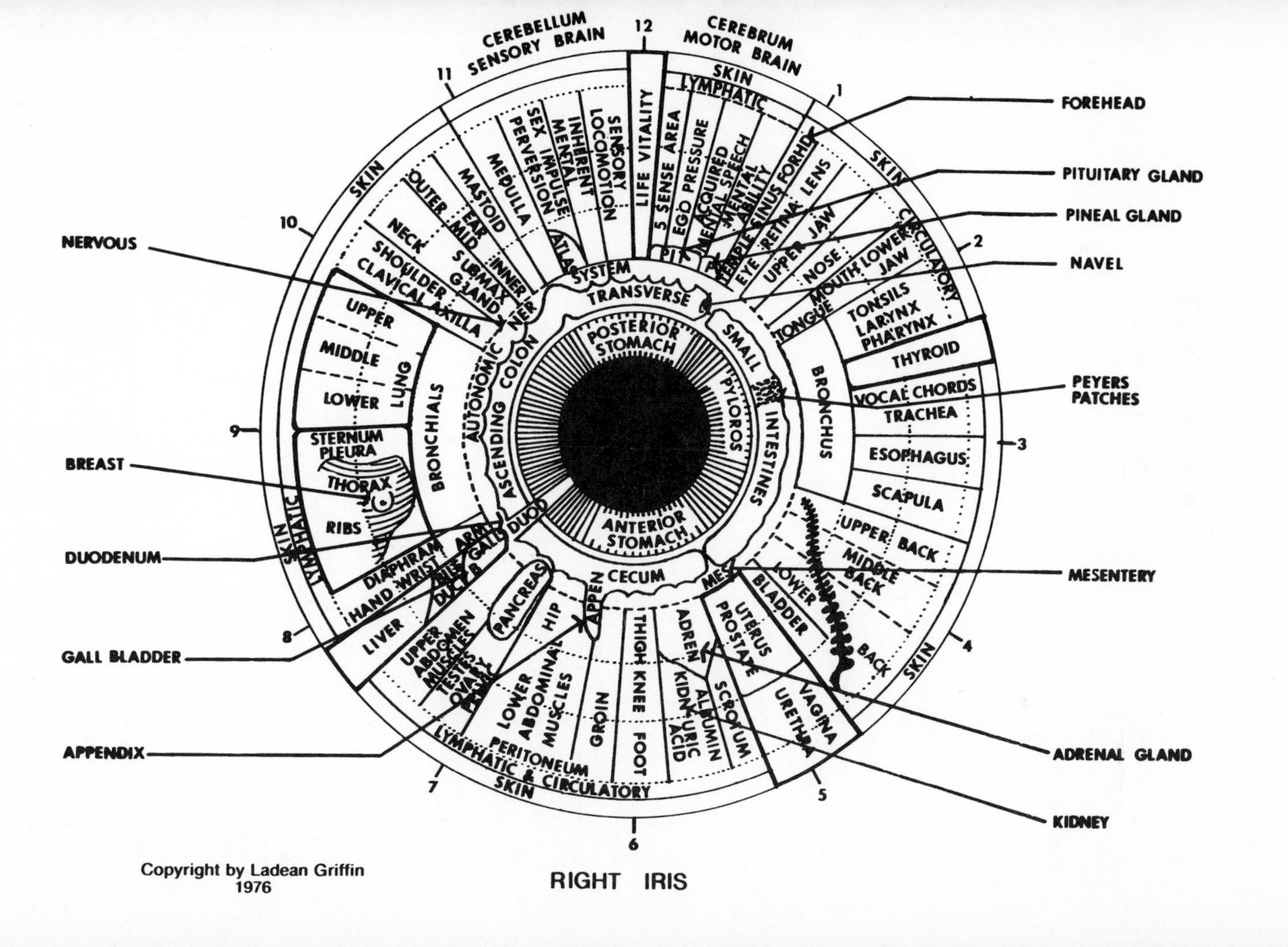
FOREHEAD
PITUITARY GLAND
PINEAL GLAND
NAVEL
PEYERS PATCHES
MESENTERY
ADRENAL GLAND
KIDNEY
NERVOUS
BREAST
DUODENUM
GALL BLADDER
APPENDIX
CEREBELLUM SENSORY BRAIN
CEREBRUM MOTOR BRAIN
12
1
2
3
4
5
6
7
8
9
10
11
SKIN
LYMPHATIC
LIFE VITALITY
5 SENSE AREA
EGO PRESSURE
ACQUIRED MENTAL SPEECH
MENTAL ABILITY
TEMPLE SINUS FORHD
EYE RETINA LENS
UPPER JAW
NOSE
MOUTH LOWER JAW
TONGUE
TONSILS LARYNX PHARYNX
CIRCULATORY
THYROID
VOCAL CHORDS
TRACHEA
ESOPHAGUS
SCAPULA
UPPER BACK
MIDDLE BACK
LOWER BACK
BRONCHUS
SMALL INTESTINES
PYLOROS
POSTERIOR STOMACH
ANTERIOR STOMACH
TRANSVERSE
SYSTEM
PIT
CECUM
MES
BLADDER
UTERUS PROSTATE
VAGINA URETHRA
SCROTUM
ADREN
ALBUMIN
KIDN URIC ACID
THIGH KNEE FOOT
GROIN
LOWER ABDOMINAL MUSCLES
PERITONEUM
LYMPHATIC & CIRCULATORY
APPEN
HIP
PANCREAS
UPPER ABDOMEN MUSCLES
TESTES OVARY
LIVER
HAND WRIST ARM
DIAPHRAM
BILE DUCT
GALL B
DUOD
ASCENDING COLON
AUTONOMIC NER
BRONCHIALS
STERNUM PLEURA
THORAX
RIBS
UPPER
MIDDLE
LOWER
LUNG
SHOULDER CLAVICAL AXILLA
NECK
SUBMAX GLAND
OUTER MID INNER EAR
MASTOID
MEDULLA
ATLAS
SEX IMPULSE PERVERSION
INHERENT MENTAL
SENSORY LOCOMOTION

RIGHT IRIS

# IRIDOLOGY CHART
by
LADEAN GRIFFIN

**Abdominal wall:**
**Ascending colon:** large intestines.
**Acquired mental speech:**
**Adrenal gland:** secreates cortin hormone allowing a person to stand daily stress and worry. Two in number, one above each kidne
**Albumin:** mucus waste.
**Anterior stomach:** front of stomach.
**Anus:** outlet of the rectum.
**Aorta:** main trunk of arterial system, (heart).
**Appendix:** overflow valve for ascending colon to prevent vac flow into small intestines and re-absorption of wastes.
**Arm:**
**Atlas:** the first cervical vertabrae, supports the head.
**Autonomic nervous system:**
**Axilla:** armpit pivot
**Back:** upper, middle, lower.
**Bile duct:**
**Bladder:**
**Brain:**
**Breast:**
**Bronchials:** tree-like tubers from windpipe to lungs, right one shorter than left.
**Cardia:** part of stomach having an effect on the heart.
**Cecum:** the first portion of large intestine located just below (illocecal valve) entrance to large intestine from small intestin
Appendix is attached to lower portion of the cecum.
**Cerebellum:** portion of brain sensory.
**Carebrum:** largest part of the brain motor.
**Clavical:** collar bone, a bone which articulates with the sternum and scapula.
**Decending colon:** left side of large intestine ending at anus.
**Diaphram:**
**Duodenum:** pylorus valve end of stomach, receives pancreatic and gallbladder bile to aid in digestion of oil.
**Ear:**
**Ego pressure:**
**Epileptic center:**
**Equilibrium:**
**Esophagus:** muscle canal from pharynx to stomach.
**Eye:**
**Foot:**
**Gallbladder:** gland secreating bile for fat digestion.
**Groin:**
**Hand:**
**Heart:**
**Hip:**
**Inherent mental:**
**Jaw:**
**Kidney:**
**Knee:**
**Larynx:** organ having to do with voice.
**Lens:**
**Life:**
**Liver:**
**Lymphatic circulation:** having to do with all lymph systems of the body, lymph capillaries, nodes, lacteals, vessels and duc
**Lung:**
**Medulla:** portion of spinal cord having to do with lower part of brain stem.
**Mental ability:**
**Mitral:** valve of the heart.
**Mastoid:** pertains to the temporal bone and sinus, lays just behind the ear opening.

**esentery:** peritoneal (skin) connective tissue between abdominal wall and intestines.
**outh:**
**vel:**
**ck:**
**se:**
**ary:**
**ncreas:** organ which secreates insulin.
**lvis:**
**nis:**
**ritoneum:** abdominal membrane over intestinal organs and lining abdominal cavity
**yers patches:** group of lymph nodules in small intestines found near its junction with colon. In high fever such as typhoid etc. the yers patches nodules undergoes hyperplasia (increase in size or ulseration) causing an inability to assimilate food sometimes rest of the life.
**arynx:** tube extending from base of scull above the sixth vertebra to esophabus.
**eal gland:** small gland attached to post part of the third ventricle of the brain. function unknown. Sometimes called the spiritual nd.
**uitary gland:** main gland of the body, having t do with growth.
**icardium:** membrane enclosing the heart and origin of the great blood vessels.
**sterior stomach:** back of stomach.
**state:** male gland.
**monary valve:** has to do with blood circulation through heart and lungs and back again for purification.
**orus:** muscle valve gatekeeper from stomach to duodenum.
**ctum:** end of the descending colon.
**ina:** receives images formed by the lens, transmits to the brain immediately, an instrument of sight.
**s:**
**pula:** shoulder blade.
**otum:** the double pouch containing the testicals and part of the spermatic cord.
**sory locomation:**
**se area:**
**perversion:**
**ulder:**
**moid:** pertains to the lower plexure of the descending colon.
**n:**
**all intestines:** where food nutrients are absorbed into the blood.
**ech:**
**een:** having to do with blood cells and toxicity of the body.
**rnum:** narrow flat bone in front called the breastbone.
**mach:**
**max gland:** neck glands under ears.
**nple forehead:**
**tes:** male reporductive glands, location in the scrotum.

**gh:**
**rax:** chest area skeletal ribcage.
**roid:** gland in throat area, secreates iodine, regulates fluids of the body.
**gue:**
**sil:** lymphatic tissue in the throat pharynx, protects from invading bacteria.
**chea:** a capillarious tube from larynx to bronchial tubes, windpipe.
**sverse colon:** upper middle portion of large intestine, colon lays between upper hip bones across behind naval.
**hra:** canal or bulb for urine to discharge from bladder to the outside.
**acid:** a crystalline acid, end product of inaccurate metabolism, seen in gout, rheumatism, etc., caused from gland imbalance cularly the adrenials.
**us:** womb.
**na:**
**al cords:**
**t:**

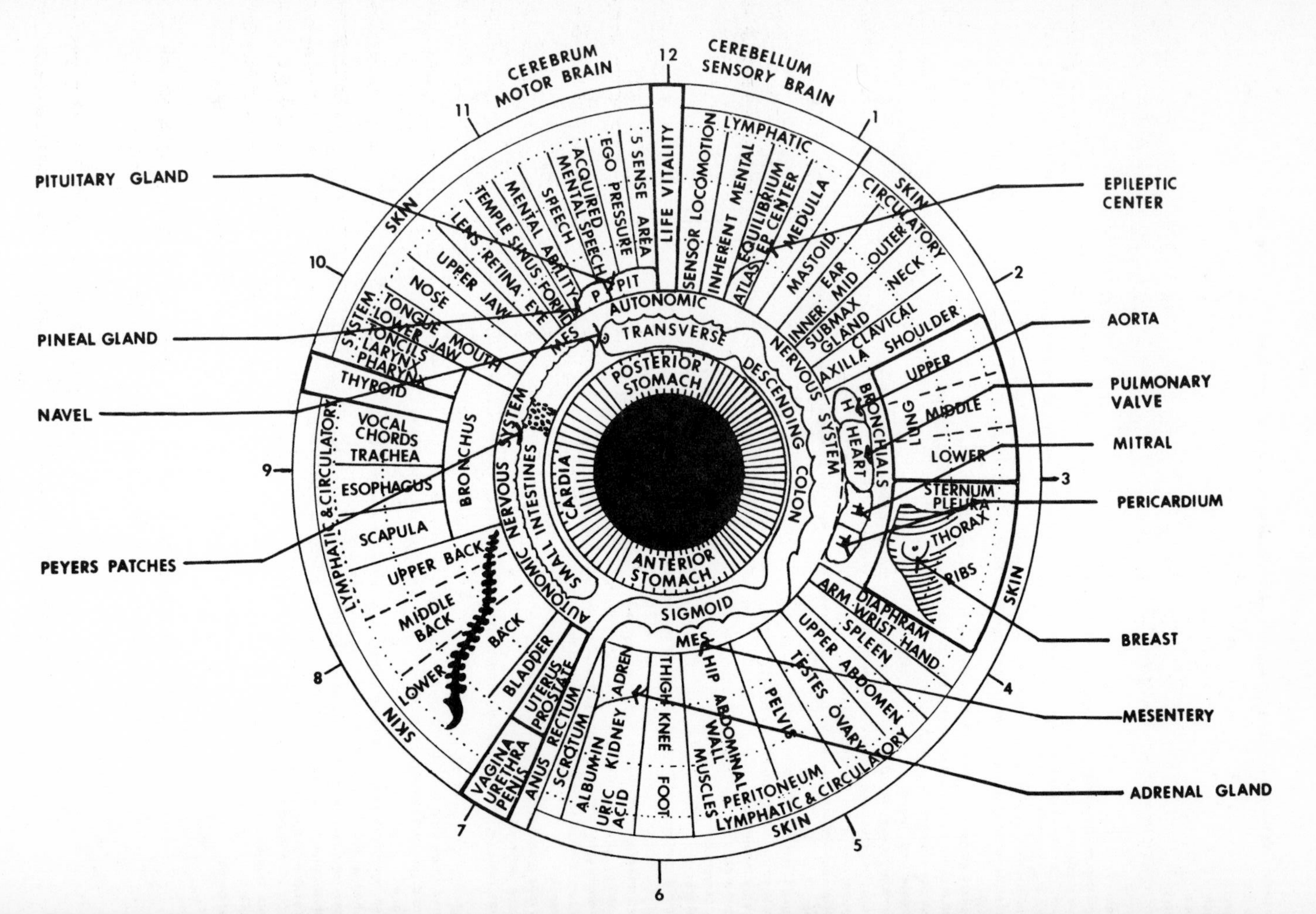

CEREBRUM MOTOR BRAIN
CEREBELLUM SENSORY BRAIN
PITUITARY GLAND
EPILEPTIC CENTER
PINEAL GLAND
AORTA
PULMONARY VALVE
NAVEL
MITRAL
PERICARDIUM
PEYERS PATCHES
BREAST
MESENTERY
ADRENAL GLAND
SKIN
LYMPHATIC & CIRCULATORY
LIFE VITALITY
5 SENSE AREA
EGO PRESSURE
ACQUIRED MENTAL SPEECH
SPEECH
MENTAL ABILITY
TEMPLE SINUS FORHD
LENS RETINA EYE
UPPER JAW
NOSE
TONGUE MOUTH
LOWER JAW
TONCILS
LARYNX
PHARYNX
SYSTEM
THYROID
VOCAL CHORDS
TRACHEA
ESOPHAGUS
SCAPULA
UPPER BACK
MIDDLE BACK
LOWER BACK
BLADDER
UTERUS
PROSTATE
ANUS
RECTUM
VAGINA
URETHRA
PENIS
SCROTUM
ALBUMIN
KIDNEY
ADREN
URIC ACID
THIGH KNEE FOOT
HIP
ABDOMINAL WALL
MUSCLES
PELVIS
PERITONEUM
TESTES OVARY
UPPER ABDOMEN
SPLEEN
ARM WRIST HAND
DIAPHRAM
RIBS
THORAX
PLEURA
STERNUM
LUNG
UPPER
MIDDLE
LOWER
BRONCHIALS
H
HEART
SHOULDER
CLAVICAL
AXILLA
SUBMAX GLAND
NECK
OUTER
MID
INNER
EAR
CIRCULATORY
MASTOID
MEDULLA
EP CENTER
EQUILIBRIUM
ATLAS
LYMPHATIC
INHERENT MENTAL
SENSOR LOCOMOTION
PIT
P
AUTONOMIC
NERVOUS SYSTEM
AUTONOMIC NERVOUS SYSTEM
TRANSVERSE
DESCENDING COLON
SIGMOID
SMALL INTESTINES
BRONCHUS
POSTERIOR STOMACH
ANTERIOR STOMACH
CARDIA
MES
1
2
3
4
5
6
7
8
9
10
11
12